Joseph Cairn Simpson

Tips and Toe-Weights

A Natural and Plain Method of Horse-Shoein

Joseph Cairn Simpson

Tips and Toe-Weights
A Natural and Plain Method of Horse-Shoein

ISBN/EAN: 9783337341381

Printed in Europe, USA, Canada, Australia, Japan

Cover: Foto ©berggeist007 / pixelio.de

More available books at **www.hansebooks.com**

A Natural and Plain Method of Horse-Shoeing;

WITH

AN APPENDIX

Action of the Race-Horse and Trotter
shown by Instantaneous Photography.
Toe and Side-Weights.

BY

JOSEPH CAIRN SIMPSON,

(Author of Horse Portraiture.)

TO LELAND STANFORD,

AS A TOKEN OF APPRECIATION OF WHAT HE HAS ACCOMPLISHED IN

MAKING KNOWN THE TRUE ACTION OF HORSES, AND THE INTEREST

HE HAS TAKEN IN IMPROVING THE STOCK OF THE PACIFIC

COAST, THIS WORK IS RESPECTFULLY DEDICATED.

INTRODUCTORY.

I present this little volume with apologies, requesting the forbearance of my readers for the want of connection, and in many cases repetitions, which are due to the desultory manner the work has been done. This is in a great measure owing to awaiting the results of experiments and taking up the subject after the lapse of long intervals. As the experiments progressed parts were written and published, and then there would arise questions which necessitated further delays. At one time I thought of rewriting the whole work, which would have afforded the opportunity to correct the lack of continuity, and then I thought it better to present it as it was, thus giving the stages as they occurred. The first chapter was published in April, 1876, and nearly eight years have been occupied in experimenting, and though confined to a small number of horses, the extension over so long a period has made the trials equivalent to a larger number of cases for a shorter time. In fact, the experiments could not have been carried out in a less space than three years, as in the case of Anteeo, a colt was taken as a subject from the first time of shoeing, when fifteen and a half months old until nearly four years of age. By taking an animal as young as Anteeo, and continuing the use of tips until so nearly matured, the test was thorough as to the effect on the feet and legs. Previous to that the differences between full shoes and tips were tried, the bearing on the relative trotting speed being the object, and the results in every case were in favor of tips, so that if it was further proven that the theory of following nature as closely as possible was the proper procedure to keep the feet and legs in the best shape, the advantages of the system could not be overrated. While offering excuses for the faults alluded to, I present with the utmost confidence my views, with confirmed belief that the system recommended is altogether the most rational method of horse-shoeing, and when the advantages are realized it will supercede that which for centuries has been known to be faulty.

I do not claim novelty in the use of tips or "lunette" shoes. They are of comparatively ancient origin. But I do claim that the manner of placing them on the feet in a great measure nullified the benefits. The plan of cutting a shoulder for the posterior portion of the tip to rest against, and thus giving the bottom of the foot a level and natural bearing, was unknown to me until I discovered it in the way which is told; and I have yet to meet a blacksmith who was acquainted with the plan until I made it public. The letters which are copied from the London *Field* were published five years after it was put in type here; and therefore I can claim priority of them. Simple as it is, I believe it to be the corner-stone of the system. When the tip is "feathered" it either has to be set with so much incline on the foot surface as to make a strain on the wall, or the toe is raised so much higher than it should be that the proper bearing is destroyed. The square shoulder is also an advantage in keeping the tip in the proper place when the nails are driven; but it makes the work of fitting much more troublesome. This is one of the reasons why blacksmiths are so averse to the setting of tips, though their greatest antipathy arises from the nearly universal determination to stick to the old ruts. Until this is overcome it will be difficult for owners to get their horses shod satisfactorily unless under their personal supervision, though eventually there will be some of the most enterprising smiths to lead the way, and others will be compelled to follow. That this difficulty will be obviated I feel confident, and in every town of any size there will be one smith who has sense enough to discover that his interests will be promoted by laboring on the side of reform. Until lately I was under the impression that shoeing-smiths would have to be educated to a higher degree, and another generation at the forge and on the floor before the work would be properly done.

I was led to this belief from the trouble there was in getting the tips made properly, and then a ten times more arduous task was to get them correctly put on. In my first experiments I made patterns of wood, and had them cast by a brass-founder; and had there been an opportunity of getting some better metal cast in the moulds, such as steel or even malleable iron, I should have adhered to the plan, but the brass wore away so rapidly that I had to change, and I was fortunate in finding a blacksmith who would forge them very nearly in accordance with the patterns furnished. Latterly I have met a blacksmith who is as enthusiastic in his advocacy of the use of tips as I am, who has put them on all kinds of horses, including those used for heavy draft, and with satisfactory results in all cases. This is Paul Friedhoffer, but as his shops are in San Francisco, the only relief to me was in being enabled to get the tips made as I wanted them, the setting having to be done by myself, as the trouble of sending horses from Oakland offset the labor. And by the way, it is doubtful if I could have continued the seven years' war, if unable to prepare the foot and nail the tips on myself. In that case I would have been compelled to superintend the job from the start to the finish, and this would have entailed the loss of more time. By doing it myself I was also enabled to see the exact state of the foot, and make changes that were found beneficial. Though the practice has made me more expert in the use of knife, rasp and file, and given me an aptitude to drive a nail where I want it to go, it is a hot job, and one that leaves a soreness of muscle which is not pleasant. Sustained, however, by the implicit faith that I was working in a good cause, that the ultimate result would bring amelioration to the animals I have been so intimately connected with for more than a quarter of a century, I have never flinched from the self-imposed task, and for the last three or four years have never lost confidence. I have listened patiently to the arguments offered by the opponents of tips, watched closely for defects in the system which the reasoning on the other side was to disclose, and which at first I thought might overturn my previous conclusions; but every succeeding year has added strength to my convictions, and given me renewed courage to adhere. Many years ago I had to select a motto to go on the records of a society, and the one chosen was "Hand fast." I have held tenaciously to the belief in the efficacy of thoroughbred blood in the trotter for fully twenty-five years, and now the granddaughter of a thoroughbred mare occupies the highest place in the record with others of the same degree of consanguinity to the royal blood close up in the calendar.

After these seven years' experience with tips, the grip has grown firmer, tightening with every successive trial, practice so fully demonstrating the correctness of the theory that nature is a correct tutor, that I hold fast as confidently as in the other illustration. In order to do so it was fortunate that I had a few horses of my own to experiment with; and though forced to offer apologies for making them so conspicuous, it was compulsory that the prominence should be given. Had there been others interested it is not likely that the tests would have been made so thorough. It is probable that a want of harmony would have prevailed, and a difference of opinion interfered.

That there has been fair success in the way of trotting speed I think is evident. The only colts I have trained in California to trot have been five of my own breeding. Three of these were foaled East, two here. Four of them have shown trials better than 2:40 when three and four years old, the other trotted in 3:02 when a yearling. With the exception of the yearling, the fastest work was done in tips, the exception being barefooted. This training has been incidental to my other avocations, and hence not as thorough as I could have wished.

The point has been raised that trotting the yearling barefooted was an admission on my part of a want of confidence in the system advocated. The reasons for trotting him barefooted are given in his history, though this is unnecessary to show the absurdity of the charge.

As light tips set in the manner described is the nearest approach to leaving the foot bare, it is manifest to the simplest understanding that the benefits of one will be shared by the other. There is so nearly the same "spring" in the foot, when the tip does not extend farther back than the point of the frog, that this valuable property is retained, and the frog-pressure is identical in both.

The appendix, I am sorry to state, is not what I anticipated to make it. Toe-weights are certainly an intricate study, or it may be better to write that the effects of weight on the foot, especially on the outer part of it, though known to be potent, the causes as yet are unexplained. At least such is my case, and I must acknowledge an ignorance which is an estoppel against any attempt to elucidate. Through the kindness of ex-Governor Stanford, however, I am enabled to give some valuable information regarding the action of the race-horse and trotter, and, with his consent, present representations of the most prominent features of the "horse in motion." Very unfortunately the copy of the work edited by Dr. Stillman, which was sent me by Governor Stanford, miscarried, and never having seen a copy I have been without the information I hoped to make use of. The first cards that were published, and a series of views arranged for the zoetrope, which were sent me by Mr. Muybridge, have given an insight of the greatest value, and, in fact, have taught me the only true knowledge I possess on the action of the race-horse. Before these instantaneous photographs were taken, the manner in which a race-horse progressed was as completely hidden as though the gallop had never been seen, and even the short explanation will be found of the greatest service, if attention is paid to the subject. Still, though it is as plain as the "writing on the wall," that a change in part of the present system of training is imperative, the pioneer who realizes the importance of benefiting from the lessons that the camera places before him has an arduous task, and the methods he employs to turn to a practical use the teachings of the Palo Alto school will excite no end of ridicule and badinage.

The time will surely come, notwithstanding the jokes and jeers, and there will be progress in this as well as other things. When that time comes, Governor Stanford will receive the thanks he has so richly earned, and the immense expenditure that was necessary to make the work complete will bring a return commensurate with the outlay. Not to the donor, as his reward will be restricted to the satisfaction of having made men wiser, and the still more satisfactory knowledge that the wisdom gained has ameliorated the condition of the horse.

It may be considered out of place, in these brief introductory remarks, to allude to a work of such magnitude as the photographing of animals in motion, the importance of the subject demanding a far more extended elucidation. Still, I cannot let the opportunity escape of reference, however brief, or pass by without proper acknowledgment of the assistance obtained from a study of the photographs. Although I had a fair knowledge of the action of the fast trotter, I was as totally ignorant of that of the race-horse as if I had never seen one gallop. Even the three cuts which are given in the first chapter of the appendix prove that former ideas of the manner in which a horse ran were as erroneous as could be, and portions of the stride are fully as grotesque as the representations given. A comparison of the cuts of parts of the racing and trotting stride will show the great difference there is in the two gaits, and even the short essay accompanying them will be found of some service. At all events, they show how much greater the strain is on the fore leg of the race-horse, irrespective of the weight on his back, and the contrast between "the last effort" and "the initial" sufficient to account for an ailing fore leg being so much worse for the race-horse than the trotter.

The appendix is not as full as I thought it would be when the opening chapters were written, nearly two years ago. Then I fancied that I had obtained a clew which would lead to satisfactory explanations of the cause for weight applied on the outside of the foot exerting such a potent influence on the fast-trotting action. The more study I gave it the more puzzling it became, and at present can only state that I am not capable of giving reasons which are at all satisfactory to myself, and conjectures would not be worth the space given or the time of the reader. It may be that future experiments and closer observation, will lead to the discovery of a key, or it is quite as likely that some trivial circumstance will prove the guide to escape from the labyrinth, and chance, as it has done heretofore, lead to the correct solution of the problem. Awaiting developments, I can only offer the excuses given, and trust that the apology will be accepted.

Before concluding, I must again refer to the want of connection and repetitions, and the necessity for bearing in mind that the publication of the various chapters was broken by long delays. The chapters were written in the following order: From Chapter 1st to 7th, inclusive, April, 1876, to 1878, then from that time until 1881, there were occasional publications, and the remainder within the last few months.

I cannot say good-by, however, without reiterating my implicit belief in the efficacy of tips, and in contradistinction from the toe-weight puzzle, there are reasons, plain and palpable, why the result of practice should be as I have found it.

JOS. CAIRN SIMPSON.

San Francisco, February, 1883.

EXPLANATORY.

A query published in the *Breeder's Gazette* a short time ago, has led me to think that it will be well to add other cuts, with an explanation, to give a perfect understanding of the system. It appears simple enough to state that the tip should be made of nearly uniform thickness, square at the ends, and the horn cut away so that the foot will have the proper bearing when the tip is set. A more perfect elucidation, however, is afforded by the cuts, and in order to show exactly the state of the foot of a four-year-old that has never worn a shoe, a few days ago Mr. Wyttenbach made the drawing from nature. The outline was obtained from placing the foot on a piece of cardboard, and carefully tracing around it. The other parts of the sketch were filled in while the colt's foot was held up, so that the artist could see as he went on with the work.

There are many points of interest to examine. In the first place it is widely different from a foot that has worn a shoe, and it also varies from one that has never been shod, if the foot has not been cared for in the same manner. The first position will be readily granted, as any one can verify it by making a comparison; the second is not so well understood, and the causes for the difference overlooked. The natural supposition would be that if a colt had run without shoes until four years old the foot would be in the proper shape; and if the animal was in an entirely natural state, roaming over the country wherever it desired, it might be so. Domestication, however, changes the course, and small fields, paddocks and stables entail a different life. In the case of Anteeo, the field was circumscribed to 200 feet by 133, with a jog that doubled the latter distance, so that 266 feet was the longest run he could take. The yard he sometimes ran in was 50x133 feet, and when not in these a medium-sized box-stall was his domicile. With no better chance to wear the horn away, the feet would have grown long at the toe, split and broken off. The heels would have got out of all proportion, and undoubtedly more or less contracted. The frog would have shrunk from lack of use, and even the inner portion of the foot between the walls been in an abnormal condition. From the time he was a few months old, his feet have been trimmed to get rid of the excess of growth, and since he was fifteen months old the front part has been protected with a tip the greater part of the time. The wall from the tip has been cut down so that it was only a trifle lower than the frog, so that the "spring" of the quarters would permit the frog to bear its due share of the weight, and no matter how ragged it became, the frog was never cut. When the new frog was ready to replace the old, there were small hanging fragments which were pulled off, though the knife was never brought into requisition further than to cut away where the tip rested, and the sole back of the tip was left intact. This, too, would exfoliate, and when a flake was so loose as to be easily removed, that would also be got rid of by prying it off. For a time I used an instrument which would cut away the horn only where the metal replaced it; the sole between the wings of the tip would also be left to exfoliate; but in order to get a true bearing, with only a knife, rasp and file, it was necessary to level that portion.

As the cut shows exactly the outline of the sole of the foot, it will be easy to determine the proportion between that and the size of the colt when the measurements are given. Anteeo is a trifle over fifteen and a half hands high, and of more than ordinary substance. His limbs are larger than usual, or, rather, it will be better stated by saying wider. As a general thing, the Electioneers have rather small feet, and I firmly believe that if Anteeo had worn shoes his would be at least half an inch narrower than they are. The width and length are nearly the same, and even a diagonal measurement, as from one heel to a point the same distance on either side of the toe, is only a trifle greater. But the most striking feature in the cut is the frog, and, doubtless, those who have obtained their knowledge of this important part of the foot of the horse from seeing those of horses which have worn shoes, or from illustrations in the books, will be surprised, and think that there is something wrong in the delineation. It is not only wide at the posterior portion, as it reaches far nearer the toe, and in place of the crevice in the center there is only a slight depression.

In a previous illustration, given on page 8, chapter 1, the only cut obtainable at the time, the representation is very faulty. In fact, this is the first instance of correct drawing of a frog that may be termed perfect, all others which I have seen being studies from feet which have become malformed.

Although the frog changes in appearance, and there is quite a difference between the old, ragged surface and the new which is ready to take its place, still the similarity of broad surfaces and elasticity is apparent. The old has served the intended purpose, doing its work until the other is ready.

When mutilated with the knife, and that mutilation accompanied by non-use, in place of this healthy growth it shrivels and becomes nearly as hard as the horn; there is a deep fissure in place of the slight depression in the middle, the longitudinal axis is shortened, and the whole is as different from the engraving as can be well imagined.

The illustrations leave little to add regarding the manner of setting the tip or applying the toe-weight which I am partial to, but inasmuch as the questions asked indicate that the proper method is not fully understood, it may be as well to be more minute in the instructions.

As has been stated, the tip is only a trifle thicker at the toe than the posterior portion, the object being to give as nearly a level bearing as possible. If there was much taper, the slope from the toe would bring a greater strain on the nails, and, consequently, a greater danger of displacement. It is also evident that the square shoulder is of material assistance in keeping the tip in place while the nails are driven, and does away with any necessity for a clip at the toe.

Many years ago I advocated driving the nails from the inner side of the wall, and the benefit of this practice has been sustained by the score of years in which I have followed it. There is not so much danger of "pricking" in setting the tip as the full shoe, owing to a greater thickness of the wall in the anterior portion of the foot; it also gives a more secure fastening, as the horn is perforated in place of the layers being divided, and a much lower hold can be taken. A horse-nail is wedge-shaped, and when driven parallel with the fibers there must be a tendency to split the layers apart, but if they are perforated there is no such risk. When the nail-holes are punched close to the edge, the nail has to be driven on a curve. At first the direction is toward the sensitive portion of the foot, and then the bevel at the point throws it outwards. Now it is evident that if the course of the nail approaches the sensitive part of the foot, there is danger, if even it is not wounded at the time of driving. Clinching the nail when it is curved throws a greater strain on the central part, forcing that part to press against the inner horn, and the concussion aggravates the tendency.

If in place of being driven on a curve the nail goes straight, the strain caused by clinching and concussion is uniform, and the straight line, at whatever angle, obviates this difficulty.

Then it is manifest that if the nail is started from the inside of the wall, and driven at any angle which will bring the point through, it never can get near the sensitive tissues. The nearest point is where it is first started, and this is so far below the quick that there cannot be any danger of wounding. In driving the nails from the inside, it is necessary to give the nail-hole the same slope outwardly, or make the hole large enough to give room for directing the point.

I prefer the latter plan, and then sink the head of the nail below the surface, which completely fills the opening. A punch is used when the nail-head is level with the shoe to drive it home, and a clinching-iron that has a projection which keeps the nail in place when it is riveted. This is when tolerably heavy tips are used; with lighter and thinner ones the head of a No. 3 nail will fill the countersunk hole. Countersinking is preferable to creasing—fullering as some smiths term it—and when the tip is so thin as to let the heads of the nails project they are filed to a level of the tip.

The countersinking cannot be done as close to the edge as a crease without giving an inward direction, and the tool with which the crease is cut is held on a bevel inclining to the inside. This, as has been shown, compels the nail being driven on a curve, first, to get sufficient "hold," and when that is done to bring it to the outside, in order to clinch it and fasten the shoe.

Anyone who will take the trouble to drive a horse-nail into a piece of wood, giving it the same curve that the smith does, twist off the point and clinch it in the same way, by splitting the wood apart after this is done, he will see how much greater the inward pressure is, and that if there is only a thin stratum between the nail and the sensitive part of the foot, there must be a pressure that will result in lameness.

When the nails are driven, in lieu of filing a notch to receive the clinch, I use a small gouge, only cutting away so much of the horn as will hold it. By following this plan, the clinch is depressed where it cannot do injury, and this without weakening the foot. As is well known, the enamel is much the strongest part, and when the edge of the rasp is used to cut a receptacle for the clinch, the groove extends from the front to the rear nail. I have known many instances where the whole side, from the clinches to the nail-holes at the bottom, was torn off with the shoe, and this could not occur when the plan recommended is followed. The toe and sides of the tip should be flush with the edge of the horn, or so near that there is only a trifle of projection to be filed away. It is better to have it exactly even, and if the gouge-cutting is properly done the clinch will be hammered uniform with the horn, so that there is no necessity for filing; and should there be a roughness, care must be taken that in removing it the horn is not marked. Every mark of the file on the enamel is an injury. It not only weakens, as it also gives a chance for moisture to penetrate, and I am thoroughly convinced that moisture is injurious and that one purpose of the enamel is to render the foot impervious to the entrance of fluids. As has been shown, the horn is composed of tubes—hairs—fastened together with an agglutinizing material, and each tube filled with a substance that gives life. The application of water weakens the adhering properties; maceration destroys; that is, long-continued soakings are prejudicial, and even washing the feet had better be dispensed with, especially when the enamel has been wounded by the rasp or file. It was also shown that driving the nails cut these tubes off, and that below the severed portion the horn is virtually dead. The low hold arising from driving the nail from the inner parts of the wall does less injury than when it takes the curving direction and a high hold, and the larger the nail the greater the damage. Four small nails (No. 3) will hold a tip of the size figured in the cut firmly in its place as long as it should be worn; and in cases where I have been negligent in resetting, the tip was not misplaced until that and the nail-heads were worn away, so that the tip could be pulled off without cutting the clinches, and yet it was retained by the slight hold which the worn nails gave.

In the cut the nail-holes are shown in order to show the manner of countersinking. It is scarcely necessary to state that when placed on the foot the nail-heads have quite a different appearance.

This description, I think, will be quite sufficient to explain the mode of setting and the reasons in brief for following that system. The toe-weights have been described in the appendix, and the cuts will complete the lesson.

NOTE.—As it might be thought that my partiality for tips and favoritism for the colt warped my judgment, I requested O. A. Hickok to make a careful and close scrutiny to see what the effect had been of constant wearing of tips on Anteeo. The appended certificate shows the result:

OAKLAND, March 29, 1883.

I have, to-day, critically examined the feet and legs of Jos. Cairn Simpson's colt, Anteeo, and hereby certify that, in my estimation, they could not be in more perfect condition.

O. A. HICKOK.

A certificate from Mr. Hickok will carry more weight with Eastern readers than that of a veterinarian.

A trainer of his skill and experience will detect anything wrong with the legs and feet, and if there is the least variation from a normal condition discover it. It is almost unnecessary to state that every individual who has seen Anteeo concurs in the views expressed by the skillful trainer and driver of trotters, and very many regard the "perfect condition" of the feet and legs as being wonderful under any circumstances.

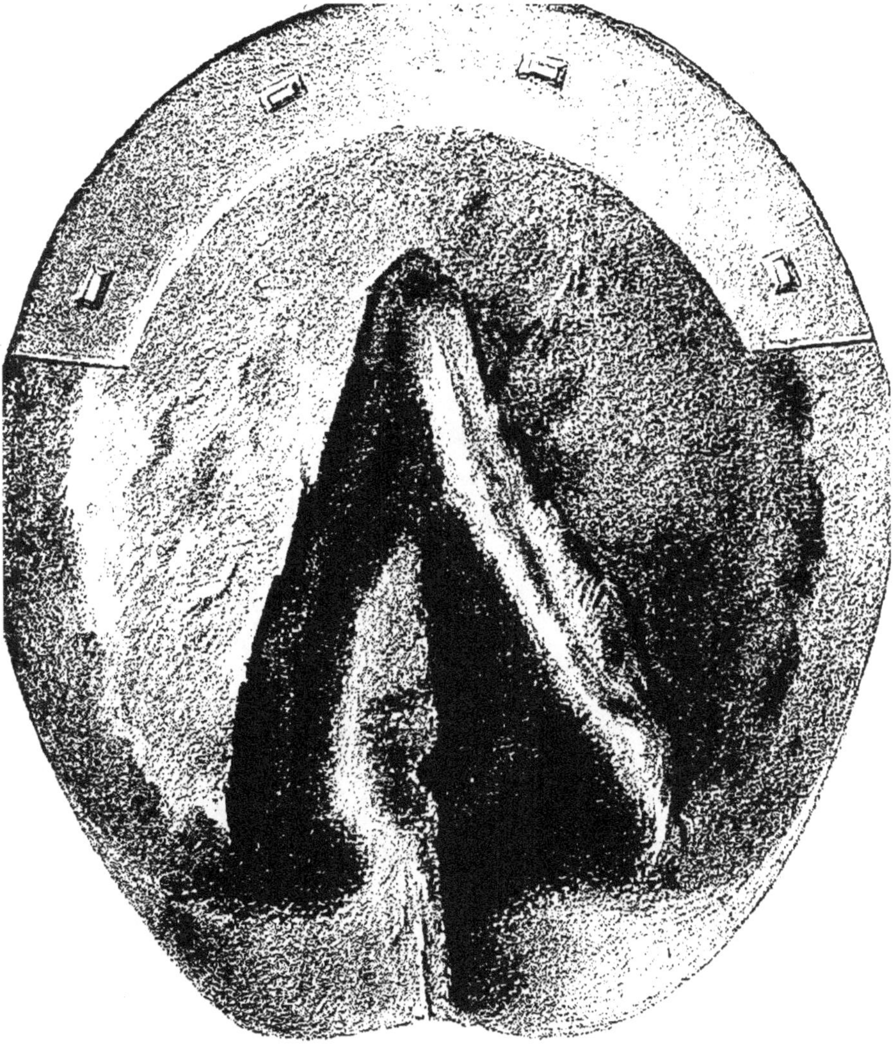

TIPS
AND
TOE-WEIGHTS.

JOS. CAIRN SIMPSON.

TIPS AND TOE-WEIGHTS.

CHAPTER I.

NECESSITY FOR A BETTER SYSTEM OF SHOEING.

The domestication of the horse, and the purposes he is used for, have necessitated the protection of the foot. In those nations the people of which could forge iron, or some kindred metal, into the proper shape, shoes of that material have been used. Different people have had various patterns, from the sheet of iron, merely perforated with a central hole, such as the Arabs put on their horses, to the elaborate articles which some smiths delight in fashioning. I am under the firm belief that all the systems, all the shapes, are more or less pernicious, and that a shoe which will preserve the natural functions of the foot has yet to be invented. Simple as the subject may appear to those who have little acquaintance with the horse, it has troubled the minds of acute observers more than any other portion of stable management, and though many have cried *Eureka*, they have been premature in their claims.

It is difficult to find two who will agree in every particular as to the proper shoeing of the horse, and now that the trotting-horse represents such an immense capital, greater efforts are constantly being made to get a pedal appendage which will meet the requirements of trainers and owners.

The art of training trotters has made rapid advancement in the last decade, and the importance of having them shod properly, and the adaptation of the shoe to the animal, has been fully acknowledged. Every trainer who gives the subject much thought, is aware of the great change in the action which can be brought about by shoeing. The stride of the "long-gaited" horse can be shortened, more or less knee-action induced, a tendency to interfere, or the wounding of the shin or knee done away with, by a change in the shoes. The race-horse, in training, wears a shoe which will weigh about forty ounces to the set. His plates, in which he runs his races, will weigh less than one-quarter of that, and the difference in his speed will be from two to four seconds to the mile in favor of the lightest. In as muscular and powerful an animal as the horse, it cannot be the few ounces of weight which effects so much, but the action, or method of running, must be favorably influenced by the wearing of the plate. If this result follows in the fast, flying gallop or run, why should not the fast trot—certainly a more artificial manner of progression—be as susceptible of the changes in shoeing? Before considering the effect of shoeing on the action, it may be as well to investigate the foot itself, and judging from the conformation of the parts, endeavor to find out if any of the present methods are rational. First, there is the wall of solid horn, compact, tenacious and altogether admirable for the natural purpose of supporting the animal, and withstanding the wear consequent upon the friction of travel. The sole, not so hard as the wall, but of different growth, with a natural provision for the surplus to drop out in flakes. The frog, still softer than the sole, and highly elastic. The wall is susceptible of dilatation and contraction, while the form of the sole and wall, aided by the commissures, is such as to permit it without injury. The bars are a sort of composite between these two, and in the sound foot there is quite a space between the frog and the posterior portion of the bars. The bars are a partial continuation of the wall to where they connect at the apex, immediately in front of the point of the frog. The commissures, or channels, between the bars and frog, are deep and angular depressions. In the natural foot there is a very slight cleft in the frog, while in the contracted, or thrushy, the frog becomes separated, as it were, in the middle, longitudinally, and a deep crevice is found. The back part of the foot is formed thus:

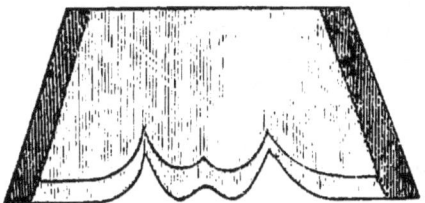

And it is evident that as the weight is thrown on the heel the quarters expand and the frog rests on the ground. The corrugated shape permits this, and when the weight is removed the elastic media brings it back to the original position. The illustration represents a section of the foot about an inch in front of where the wall and bars form an acute angle. The bar forms a curve sweeping from the point of the frog to the junction above alluded to, and the fissure between it and the frog is the deepest at the point which the cut delineates. The wall is much stronger at the heel than what is termed the quarters, and only slightly thinner than at the toe. This is necessary to withstand the concussion, which is the greatest at the heel, and the thinner quarters admit of more freedom of motion. The expansion is not confined to the ground surface—the whole foot, from the coronary ligament to the bottom, expands and contracts as the weight is thrown upon the limb, and taken from it when the animal raises his foot to thrust it forward. In the unshod, natural foot, there is nothing to prevent this freedom of motion; it is unfettered, and performs its functions without hindrance. In the gallop and the fast trot the body is hurled through the air, there being a time when all the feet are off the ground. The eye detects this quite readily when the horse is running, but it is extremely difficult to discern it in the trot. In the gallop the feet touch the ground in regular succession, followed by a bound, and the ear is cognizant of the fact, if the eye fails to be convinced. The sound of the gallop can be closely imitated by resting the thumb on a table and striking rapidly first with the little finger, second, with the one next to it, then the middle and fore-finger; the interval, before the little finger can be brought down again, corresponding to the bound or leap. Only one foot touches the ground at a time, when, of course, the whole weight must rest upon that foot, although it is quickly relieved by the other. As at least two strides are made in a second, and as the longer time is occupied by the bound, the contact of the foot with the ground being a small fraction of the time, the duration lasting while the body is carried far enough forward to cause it to be taken up. One of the fore feet supports the whole of the weight of horse and rider, while the body is moved over it, until it has to leave the ground; then the bound follows, and the hind foot strikes eighty-six inches beyond the imprint of the front foot.

In trotting, the feet are moved diagonally, and so nearly at the same instant that the ear also fails to separate the footfalls, and the sound of the feet of a square trotter, at a twenty gait, has a regular cadence like the ticking of a pendulum. The foot-prints of a race-horse show plainly the flight of the body through the air, and can be pictured thus:

4   3   2   1   4   3   2   1

We will suppose the figure 1 to be the right fore foot, 2 the left fore foot, 3 the right hind foot, and 4 the other hind foot. The length of the stride is manifestly from where the foot left the ground until it struck it again. In race-horses this will vary from 18 to 25 feet. The impress which the trotter leaves is more nearly equidistant, thus:

4   3   2   1   4   3   2   1

and from 17 to 22 feet is the length of the strides of the fast trotters. Adopting the same simile, from 1 to 1 is the space measured by each stride, and it is just as evident that to enable the animal to strike the ground twenty feet in advance of the position the same foot occupied, the body must be thrown through the air when all the feet are off the ground. It is also necessary that when the foot strikes it must be thrust forward as far as it can be and sustain the equilibrium, and hence the heel must receive the first shock of the contact. *The whole concussion is upon the heel; nearly the whole of the friction upon the toe.* How admirably the foot is fashioned to sustain this concussion and friction is apparent to any one who will give it careful thought. When the horse is at rest, nearly two-thirds of the weight is supported by the fore feet, and the sound animal does not attempt to relieve them from the portion of the weight it is their duty to sustain. In that case there would be a moderate expansion of the foot. When he walks, three of the feet are encumbered by the weight, while the fourth is moved forward, and the foot which is without a load is contracted, the others more expanded, than when the animal was standing. In the trot of the ordinary work-horse, the pace is slow and the stride short, the hind foot dropping into the track of the front. As velocity increases the force, the slow trot is more trying than the walk, while the great rate of the fast trotter, and the still faster flight of the race-horse, entails the greatest possible concussion, the greatest jar to the feet, and the greatest strain upon their mechanism. When the foot is free to dilate, so as to avail itself of the benefit of the spring of the wall from the toe back, and the further breaking of the jar by the soft and yielding pad of the frog, the great strain of even the gallop, augmented by the weight of the rider, can be sustained. The unshod foot may wear away from the friction of hard roads until the protecting covering is removed and the sensitive portion laid bare, when the horse becomes lame, but the evils arising from concussion are unknown. In the last year I have inquired of at least a hundred gentleman who were familiar with the horse in early times in California, when shoeing of saddle-horses was comparatively unknown, if they had ever known quarter-cracks in this class of horses? "Never," was the unanimous response. Were corns frequent, or bruises of the sole of common occurrence? The answer would be: "I never knew an unshod horse to have corns, and the only drawback was the wearing away of the horn at the toe until they became lame, when the animal was turned out, and a few weeks, especially in the rainy season, furnished the remedy." That shoeing is the cause of quarter-cracks and corns is absolutely proven by the absence of these difficulties, when the animal remains without these artificial appendages, and I hope to show in this essay that the reasons for the injury are apparent after due consideration, and, perhaps, suggest a remedy.

First, as to cracks: It is well known that the fissure starts from the coronet, in some cases being less than three-quarters of an inch in length. The shoe preventing the expansion of the lower part of the foot, the upper bursts the shell at the thinnest portion, as the weight is virtually thrown upon it. Unquestionably the horn in some horses is more disposed to crack than others, but the feet which have the greatest inclination to separation of the horn are those which are the nearest vertical, and in which the horn is usually the thickest. If even the nailing does not come further back than the quarters, the friction between the wall and the shoe soon wears a channel which restrains the heels from opening, and this, aided by the greater concussion upon the wall which the shoe imparts, splits the fibres apart. In the foot which is unshod, the wall, the bars and the frog unite in supporting the weight, while the spring of the heels and the elasticity of the frog break the jar. The common practice of smiths is to thicken the shoe at the heel and make the web narrower, which renders it impossible for the bars and frog to perform their natural functions. The whole force of the blow, for blow it assuredly is, comes upon the wall and the junction of the wall and sole; and confining that which should expand, if there is any brittleness it gives away. We have heard it stated that the cause of quarter-cracks was the contraction of the hoof, forcing upward the coffin bone until the wings wore the shell so thin that it split from the erosion. So absurd is this theory that it is difficult to believe that it could ever have been entertained by any one who had given the horse's foot the least attention; but my informant hinted that it was part of the teachings of a man who claims to have made discoveries that are of vast importance, and which have been sustained by men of wisdom and acumen. Inasmuch as part of the teachings of this gentleman are covered by the seal of secrecy, it may be that this is one of his esoterical points. It is obvious, however, that the wearing of the horn by the bone would result in something far more serious than quarter-cracks, and before the grinding of the wall had reached the point of cracking there would not be any foot to split. That the shoe confines the foot is easily proven by comparing the foot of a horse which has never worn shoes and one with these appendages. The first is susceptible of being forced to quite a distance by a slight pull; the other cannot be moved a hair's breadth by the strongest effort of the hand. Take the unshod foot between the knees in the same manner a blacksmith holds it when preparing it for or nailing on the shoe, and grasping the heels, with the thumbs bearing on the commissures, and the yielding is not only felt, but is apparent to the eye. Grasp the foot higher up and the fact of the contraction and expansion of the upper part of the foot is also ascertained. Taking the shod foot, especially one that has been shod for years, and while the lower motion is entirely gone, that below the coronet has been very much lessened. In the natural foot the frog is wide, and as was remarked before, the cleft of it is shallow. In the shod one the frog has dwindled to one-half its original proportion, is hard, without elasticity, and in many cases the blade of a knife can be run into the cleft an inch or more. Another great change will be observed. The foot, in its normal condition, is wider than it is long. In the artificial state, induced by shoeing, it is much longer than wide. The narrowing or contraction of the heel has lessened the transverse measurement and increased the longitudinal. The absence of corns in the unshod foot proves that the concussion is the most violent between the shoe and the foot, for if these troublesome things resulted from contact with the ground, the want of shoes would be favorable for their production.

The following cut shows the ground surface of the foot, and the peculiarities of formation, which the previous illustration only partially represented, are here passably portrayed. It is very difficult to give a draughtsman a proper idea of the shape of the natural foot, unless the animal can be present. Few of the drawings of the horse's foot which are given in the books are correct in showing the natural foot, the artists evidently having taken their "studies" from specimens which had been changed by shoeing. The frog is usually represented as much narrower than when in a normal condition, and the commissures also more contracted and with different outlines. The widest part of these channels is about midway between the point of the frog and the heel, the outward line having quite a curvature, the inner, or that side which is formed by the frog, being nearly straight. The commissures are also the deepest at this point, so that there is an arched line running from the point of the frog to the heel. It is evident that the dilatation, when the weight is thrown on the foot, is made easy by this formation, and consequently the contraction, when the weight is removed, is imperative.

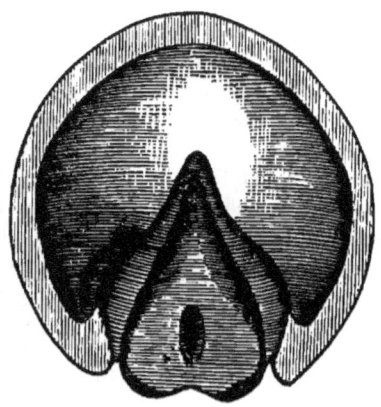

That these are natural and essential functions is almost discoverable from the study of the cut, without further knowledge. The thicker horn at the toe, which is diminished at the quarter and again increased at the heel, proves that the greatest wear will be at these points; or rather, that these extremities are the most thoroughly protected by the natural growth, while the more yielding quarters afford the necessary elasticity. The deposit of horn is greater at the toe and heel, which gives the natural foot something of the shape of the human pedal. This does away with the violent concussion at the weakest part of the foot, throwing the work of sustaining the shock of rapid locomotion on the heel and the frog, while the toe sustains the friction of retaining the hold of the ground until the foot be again elevated. While the sole performs its part in sustaining the weight, that part is very slight on even, soft ground.

Experiments have proven that the sole may be entirely removed without the bones descending from their position, the coronary ligament, and the tendons which run between the navicular bone and the sensitive frog, being sufficient to carry the weight. Veterinarians have argued that the descent of the sole proves that at times there is great weight thrown on this part of the foot, and point to a large and wide-heeled foot as the most likely to suffer in this way. There may be other causes which conduce to this state, and these causes we will consider hereafter. To digress, however, in this connection it may be as well to describe another portion of the foot which has a great deal to do in sustaining the weight, and that is the *laminæ*, or thin plates which interlace from the wall and the coffin bone. There are about five hundred of these plates, and they come together much the same as would the leaves of two books, if they were locked at their edges. With even a slight pressure from the sides, it will be found that it takes considerable force to push them apart; and the five hundred *laminæ*, thin as they are, aid very materially not only in sustaining the weight, but also in breaking the jar. These *laminæ* permit the dilatation and contraction without injurious results, as would be the case were the interior surface of the wall and the outward portion of the bones smooth. In the latter case there would be nothing to hold them in position while the foot was expanded, and there would be a space between.

The cut in the previous article showed the section of the foot an inch in advance of the heel and where the commissures are the deepest, but as that becomes shallower there is less depth of horn to move, and less is required. The elasticity of the upper portion of the foot is the greatest at the back part of the coronary ligament, and the equilibrium of expansion is thus retained.

Miles, in his essay on horse-shoeing, wrote: "Unless the nail-holes are placed so that the foot can expand, it must in the end become unsound." "Frank Forester" thought so highly of his treatise that he embodied it in his work, and prefaced it with unqualified praise, and his opinion was coincided in by the best informed horsemen of the time. He expected to overcome the difficulty by placing three nails on the outside of the foot, and two on the inside, near the toe. He also claimed that "The portion of the hoof that expands the most is the inner quarter and heel." The horn on the inner quarter is slightly thinner than on the outside, but there is a greater concussion on the outside, which makes the thicker horn give more. But it is recently that the expansion of the foot has been considered unnecessary, and that only by a few.

The cuts show as conclusively as that twice two are four, that there are natural provisions for expansion, and as the "good mother" is usually correct in her doings, we will accept it as the proper thing to follow. "Contraction" is claimed by many to be the result of disease; others contend that contraction produces disease of the foot. It is immaterial which is correct so far as regards the consideration of the evils of the present system of shoeing, as the fact of it being apparent in nearly every horse which has been shod for a series of years is sufficient to warn us that something must be done to counteract this tendency.

Mr. Miles was not alone in advising that the nailing should be confined to the anterior of the foot, Mr. Coleman, Professor of the Royal Veterinary College, having preceded him in this recommendation. He taught in his lectures to the students, "that, for a good natural foot, all that is required by way of a shoe is to guard the crust by a small and narrow piece of iron, which should be attached principally at the toe." Mr. Bracey Clark, one of the foremost writers on the pathology of the horse's foot, noted the effect of shoeing was usually to contract the foot, and his instructions were to leave the sole, bars and frog in the natural state ; and having seen ill effects follow the non-nailing at the heels, from sand getting between the shoe and foot, and more concussion, he nailed back to the posterior extremity. It is useless to multiply the evidence of those who join in denouncing the ordinary system of shoeing, and who have proven the injuries which have followed. It would be far beyond the limits of this essay to note the various opinions which have been given, and to record the acrimonious debates between partisans of the different schools. Treatment as opposite as the antipodes has been authoritatively recommended, and between these multitudes of counselors the poor horse has sadly suffered. The horse must be shod is taken as an axiom, but yet we find some of the most intelligent of the present trainers of race-horses galloping them without any protection to the foot. On the soft dirt of a carefully kept race-course there is little, if any, necessity for protecting the hoof; but in California, where the courses are hard, the wearing away of the toe has to be guarded against.

It has been deemed essential that something should intervene between the foot and the ground "to break the jar" consequent on the weight of the animal coming with such force upon the extremities, and the harder the track, the opinion of a majority of race-horse trainers has been that a heavier shoe was required. I do not think it will need much argument to show that the "jar" is greater the heavier the shoe, and also that there is greater strain on the limbs. The iron is unyielding, and without elasticity, in the sense we call the foot of the horse elastic. Weight induces higher action and a longer stride, and hence there must be greater force in the blow when the foot strikes the ground. The hard metal, while it protects the horn from being shattered by guarding the edge, impinges on the angle made by the bars and wall, resulting in corns and bruises. Unfortunately, that is not the greatest injury. The expansion is hampered, if not completely stopped; the sole and frog are not permitted to bear their due proportion of the weight; the latter becomes atrophied, and the former loses its natural properties.

Great masses of iron nailed to the horse's foot have been the weakness of shoeing-smiths of the last two centuries, and Mr. Clark gives illustrations of the forms which many then held to be correct. The web covered nearly the whole sole, there being only a small circular opening at the point of the frog, and a triangular one under it. Calkins were turned at the heel, and this ponderous body was fastened to the foot by fourteen or fifteen nails. Horses which had gone lame while shod with this *nightmare* of a shoe, recovered with the narrow, concaved one which Mr. Clark used, notwithstanding the foot was fettered from the toe to the heel. The Duke of Newcastle saw that for horses of the "mannage," and for those which were used for hunting and hacks, there was an impropriety in thus loading them, and recommended a lighter shoe. For horses of heavy draught it is still held to be necessary to have iron, in mass according to, and being in harmony with, the size of the animal, claiming that on the pavements a lighter shoe would soon wear out. As the wear is nearly all confined to the toe, the greater part of this could be dispensed with.

But there is one class of horses which must have weight on their feet, or some contrivance which will have the same effect on the action as weight, and this is the fast trotter, or, at least, a great many of the very fastest. Thus, a little mare like May Queen has to wear a shoe weighing twenty-four ounces on each fore foot, and Jenny had a still heavier incumbrance, with a "toe-weight" of like ponderosity added. That these two mares could make "a record" of 2:20 and 2:22 while thus loaded, proves that the benefit to the action offsets the disadvantages. Jenny made one brilliant season, and then had to be retired until her legs recovered from the strain. Nettie, another first-class animal which has to be burdened in the same manner, has several times been troubled " with a leg," and these appliances have been condemned on account of the greater liability to the tendons being injured when they were used. When the time for the consideration of the toe-weight more appropriately comes, I will endeavor to show that it is not a necessary sequence to their use, but results from a wrong understanding of how they should be applied. If the proper action can be obtained, and the foot properly protected, I imagine few will disagree with me, that the lighter the weight of the shoe the better it will be for the animal wearing it. The pedestrian, when he has a distance to go at his best pace—say a run of a mile—wears shoes which are only a few ounces in weight, with spikes to prevent him losing ground by slipping. If he is going to walk a long journey, he wears a thick-soled shoe with light uppers. This has been offered as an argument favoring heavy shoes on horses, but there is not the least analogy between the two. The Indian performs great feats with only moccasins to protect his feet; those of the white man are tender from always having a stronger protection to guard them, and have become more sensitive to pressure and more liable to bruises. But there is scarcely anything parallel in the human and equine foot.

The man has the whole bearing from the *os calcis* to the *metatarsals;* the horse has a point in comparison, the part corresponding to the human heel forming the point of the hock, while the foot has only three comparatively small bones to sustain the whole of the shock. Again, these bones are enclosed in a box with a very slight interposition, and when this box becomes narrowed it is something like the old instrument of torture, the iron boot, with wedges and screws to compress the enclosed limb. Those who have worn a tight-fitting boot know what the pain is without the addition of wedge and screw.

An acquaintance who was fond of horses, and paid a good deal of attention to them, had purchased a new pair of hunting boots, and spent several days on the large islands in the upper Mississippi, duck-shooting. They had stiff quarters, and the heels were rather too narrow for his feet. He suffered a good deal while tramping over the wet ground, but the excitement of the sport sustained him, and though he "hobbled" along in the morning, once fairly warmed to his work, the pain was unheeded when the whirr of the rapid flight of the teal or mallard was heard. Even when lying by the camp-fire with the obnoxious boots removed, there was a good deal of suffering; but the solace of some of "Billy G's" doubled-distilled and a pipe of fragrant kinnikinick made it endurable. When he got home he was nearly tired out, and the bed was welcomed only as one who has camped for a week on a Mississippi island can greet it. Rip Van Winkle never slept sounder, until the burning, compressed *quarters* partially awoke him. He fancied his pedal extremities had changed, and in lieu of a fairly-shaped "No. 7" the hoofs of a horse were substituted. His moanings awoke his wife, and her inquiry of "What is the matter?" elicited the response, "If you do not send me to the shop and have my shoes pulled off, my feet will be ruined."

"What on earth do you mean by having your shoes pulled off at the shop?" was the next query.

"Why," he replied, "my feet are badly contracted, and these shoes are holding the quarters as though they were in a vise, and if they are not removed, and I am allowed to run barefoot for a while, my feet will be ruined entirely."

Thoroughly alarmed, thinking he had become insane, his wife jumped out of bed, turned up the night-lamp, and anxiously looked for some further token of his lunacy, and it was only after taking hold of his feet and thrusting them from under the bed-clothes so that he could see them, that he was convinced that it was a phantasy.

Compassionate before, ever after that hallucination the best care was taken of the feet of his horses. He has made the subject of shoeing a study, and has suggested many things to ameliorate the ill-effects arising therefrom.

CHAPTER II.

Guards Against Concussion—An Elastic Shoe—Stonehenge on Tips—Etc.

Pressure on the sole causes pain—that is, a degree of pressure which is less than might be inflicted by a stone striking the sole at the junction of the wall and sole. This is proved by pulling off a shoe with pincers, for if it does not yield readily the horse will flinch when the "purchase" is on the sole. The more central portion is not so sensitive, and the anterior part is so strongly formed that heavy blows do not produce pain. In the natural, unshod foot the wear is so slight at the quarters and heel that the horn of the wall and bars projects enough to guard the sensitive parts, and a further guarantee against injury is the elasticity of the natural sole and the spring of the frog. No matter how heavy the shoe which is used, there must be violent concussion when the foot strikes the ground, when the natural spring is rendered unavailable. The heavy freight-wagons of the cities, which have springs under the bed sustaining the load, are found to last much longer than those which are without these appliances to moderate the shocks of the pavements. The wheels, axles, and even the tires do not wear in the same proportion, and the draft is also lessened. Without springs the jar is sudden, and the impinging force has nothing to break it. Without being accurately a parallel case to the foot of the horse, there is a good deal of analogy between them. The living tissue, while it is more sensitive, repairs itself when permitted to rest; recuperates when the cause of the injury is removed. A shoe with the web so wide that it would cover the bars, and press equally on them and the wall, would be preferable to one which brought the whole of the bearing on the outside crust. The former would distribute the blow over a wider surface, while the latter would confine it to the part most keenly sensitive of any portion of the foot.

The wall and sole are fastened together by agglutination, and maceration will separate them. Corns are produced at the junction and in the angles which the bars form ; and bruises, at times, result in serious diseases. From the bruise the living tissues are affected and pus is created. This cannot work through the horny covering, and the suppuration has to find vent at the coronet. Oftentimes a horse, after being very lame, recovers when the opening takes place, and the general verdict is that a gravel has been embedded in the foot and worked through, as all foreign substances have to take this upward direction before they can be got rid of. Bruise, or a suppurating corn caused by the impact of the shoe, is the true explanation of the trouble. Among several other shoes which I have experimented with was one formed with two plates, between which was a stratum of rubber. Finding that the elastic material placed between the shoe and the foot was liable to misplacement, the plan was adopted of fixing it by two plates. The trial was satisfactory, and the benefit of this shoe was evident. Acting like the spring under the load on the wagon or car, it moderated the concussion and broke the jar. But the difficulty remained, the trouble arising from the stress which was put on the wall.

To get the amount of metal which would withstand the wear of macadamized roads, necessarily made this shoe thick, and the foot was raised too high from the ground. And finding that the concussion was only detrimental at the posterior portion of the foot, a shoe was made like the cut on the following page.

The ground surface, which is shown by Figure 1, was made wide at the heels, giving a full bearing on the wall and bars, and the elastic material was confined to this part of the shoe as shown by Fig. 2. The covering of all of the foot, save the frog and so much of the sole as is anterior to it gave—as nearly as possible—a natural bearing, and by extending the surface the elastic material was increased in volume.

This shoe was very satisfactory, and obviated many of the difficulties. It was an endeavor to approximate to the teachings of nature, and, like many others, I was ready to shout with the Greek philosopher, "I have found it!" It may prove that this is a pattern shoe, and the necessities of domestication require something of the kind. But so far I am convinced that the principle is wrong, and the simple plan is to follow nature literally so far as artificial requirements

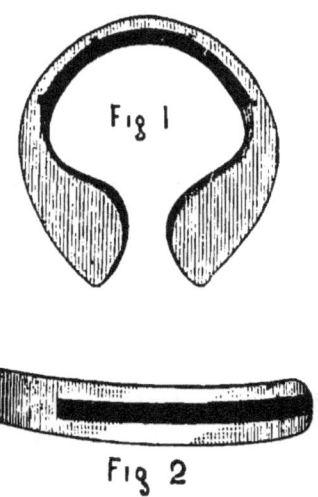

will permit. Before giving the results of late experiments, I will call attention to the following extract by Mr. J. H. Walsh, F. R. C. S. ("Stonehenge"):

"Before proceeding to describe the various methods adopted in shoeing the horse, it will be well to consider whether it is necessary to protect his feet in this way at all. This has been doubted by many, and an attempt has recently been made by Lieutenant Perry to prove that even in this country a horse can work on our roads unshod. His opinion, and that of the few who coincide with him is, that if the foot is gradually accustomed to the friction of the road, it will secrete a stronger horn, and throw it out more rapidly, so that it will bear the enormous wear and tear which its use on our macadamized road entails upon this organ. This argument is supported by numberless instances abroad, in which horses are used without shoes; but it does not follow that because they will bear the friction and blows incidental to one kind of surface, a different one will not lame them. Every experiment which has been made in this country of working horses unshod has turned out a failure, and in Lieutenant Perry's case the mare on which he tried the plan became so sore that his commanding officer interposed to prevent a further continuance of the trial. It can only therefore be considered conclusive by those who are willing to take the opinion of a Colonel of Engineers as opposed to a subaltern officer—which is the position in which this single experiment stands. Every horseman knows that without a gradual seasoning there is no doubt about the foot being too weak to stand the wear of the road, and therefore unless the trial is made under every advantage, it goes for nothing; and the mere fact that a horse, after losing a shoe, can hardly be taken home without breaking his foot, proves nothing, because it may be alleged that the same animal, if left unshod, would in course of time secrete a horn so tough and hard that it would be capable of bearing any amount of friction. Judging from those cases in which I have seen the plan partially tried, with tips instead of full shoes, I believe it is impossible to make it succeed with high-actioned horses on our roads during the summer seasons, for even with that protection the heels and frog become very thin, and I am satisfied that the toes, if unprotected, would wear or break away to the quick in a very short time. Whether it is possible to work any horse, possessing an average foot, with tips only, on our roads, I am by no means prepared to say, but that some horses can do so I know from positive experience. The heels wear thin, but do not become bruised, and the horny matter of the frog is renewed as fast as it is required. Undoubtedly the toes, when unshod, are much more exposed to injury than the heels, especially in those horses whose action is inclined to make that part touch the ground first, for there is a tendency to break as well as wear away. It is also an admitted fact, that many thousands of horses are annually lamed by the present system, and therefore I should much like the system of shoeing with tips tried on a large scale. The question is, whether those horses who bring their heels down first would be able to bear the bruising of the frog which this action causes; and if not, it would always be a doubtful point which must be left to the discretion of the smith, whether every individual horse should be shod in one way or the other. Unless, therefore, tips could be used in a vast majority of cases, I do not expect much good from their introduction."

It is evident from the quotation, that "Stonehenge" thought highly of tips from a theoretical standpoint, but was afraid to recommend their adoption until they had been further tested by a more general use. The only drawback, he said, was the wearing of the heel and frog, the liability to bruises not being deemed more likely than with the ordinary shoes.

His qualifying sentence of "unless, therefore, tips could be used in a vast majority of cases, I do not expect much good from their introduction," is not sound. If tips prove greatly superior for the fast trotter, the roadster, the race-horse in training, and other horses that have light, fast work, it is not necessary that they should be worn on the coach or heavy draught-horses.

The advantages of having the quarters unfettered, and a proper pressure on the sole and frog, are apparent to any one who has given the requisite thought to the anatomy of the foot and requirements of nature. If some of the purposes which the domestication of the horse has entailed be antagonistical to the wearing of tips, and compel the use of the full shoe, the trouble will be with that class of duties. In our opinion, however, it has not been proven that such a state exists, and that a horse with a sound foot will work with them in any situation.

The ordinary shoe, with high, sharp calkins, would seem to be indispensable on frozen, ice-covered roads. With a low, keen projection on the toe of the fore foot, and the hind foot shod in the usual manner, there would be little, if any, danger of slipping on the ice. The frog has quite an adhesive property when permitted to grow; it retains its full power by constant usage, and the toe would be the only point which would be likely to slip.

The hind feet have a double duty to perform, viz.: propelling the body, and bringing it to a stop—at times suddenly. The greatest strain is upon the toe in the first and the heel in the last. Fortunately, the hind foot, when compared with the front, has almost a complete immunity from diseases which are so common in the anterior supports, and while there would be advantages in having this as free as possible, if necessary it can be made to wear whatever is required to guard against slipping, and make its powers of propulsion and stopping more effective.

The pavements of a city are not so wearing to the foot as those roadways which are covered with gravel or broken stones. Calkins do not protect the feet on cobble-stones or the Belgian blocks, and the wear to the heels and frog upon them would not equal the natural growth.

Placing a shoe on the draught-horse's foot, which is elevated from an inch to an inch and a half by the high "corks" which the owner directs the blacksmith to forge, takes away a great deal of his power. It places him at a disadvantage to use his strength, and makes a strain upon the wall of the foot. Were the same danger to exist from this cause as results from the concussion attending a high rate of speed, the feet of the work-horses would be in a sad state after a short time. The slow pace moderates the concussion, and the damage is not so apparent. Horses which have to go fast are the ones which absolutely require an improvement in the manner of shoeing, and though confident that the best for them would also be the best for the work-horse, the consideration will be restricted to the class I have designated.

The various shoes which have been in use for the past twenty years have some radical defect, owing to a prejudice which all have entertained for complete protection of the horse's foot. Because men would cripple when walking barefoot over a comparatively smooth surface, it was supposed that the horse would be subject to the same inconvenience, and something between the foot and the ground was held essential for their comfort. Although the physical formation of the man and animal be something alike, the feet are opposite as can be. Man has the whole bearing from the *os calcis* to the end of the toes, and the metatarsals are arched so that there is a spring from the heels to the toe. The only protection which nature has given is the disposition of the skin on the sole of the foot to thicken by usage and become much harder when exposed. Still, it is only skin, and quite pliant, even when the thickest. As I have remarked before, the horse has three small bones encased in horn, and the *os calcis*, in place of touching the ground, is elevated from twenty-two to twenty-five inches above it. Nature has given ample protection from usual occurrences in the thick, hard wall, and quickly-growing sole and frog.

The foot of a man can be encased in a boot or shoe, which scarcely interferes with the greatest freedom of action, and yet, when called upon for rapid exertion, the usual covering is found to be detrimental, and a lighter shoe, and one which does not hamper, takes the place of the ordinary one.

The first articles illustrated how nature had provided for the freedom of the horse's foot, and how completely it was fettered when a band of iron was nailed around it. Another comparison will be more *apropos* than that between man and the horse.

The fastest animal next to the horse (and some contend that he is the faster of the two) is the English greyhound. His foot is very different from the horse, in having quite a spring between the heel and the toes, which comes from the knuckles forming an arch. The pad is protected with a sole like the indurated skin of the human foot, but not nearly so hard as the layers of horn of the sole of the horse. This pad at times becomes worn, so that it has to be artificially protected. It is obvious that this protection will, in a measure, confine the toes, and the animals wearing them have nothing like their usual speed.

When the dog is at rest, the toes are drawn together, and the foot covers a small space; but when in the full force of the gallop, covering twelve to fifteen feet in his stride, the toes are spread apart, and the imprint is nearly double that of the foot when raised. The greyhound is lighter in proportion to his size than the horse, and consequently there is nothing like the concussion of the heavier animal, even without a rider. A weighted collar upon the neck will handicap a fleet dog to the level of a slower, and a few ounces in this collar has a great effect.

To return to the difference between shoes and plates on a racehorse, and the effect of the latter in increasing the speed: it is manifest that it is not the trifling difference in weight which causes it. It is generally supposed that the plate induces lower knee-action, and that increases the speed. Were that the case, those horses which have the least knee-action would be the fastest, which does not follow; but the thin strip of iron more readily springs with the foot, and with the diminished thickness permits the frog and sole to come to the ground and moderate the jar. This gives something of the freedom of the spreading foot of the greyhound; the horse lengthens his stride, and "gathering" more rapidly, the action is better.

But there is a drawback to the plates, and that is the greater liability for the foot to become sore from the pressure of the narrow iron on the wall, and bringing the whole force of the concussion on the part that is the most sensitive. Trainers and jockeys talk about the hard track "burning" the horse's feet, when they notice the shortening of the stride and the endeavor to relieve them by "changing feet," when the injury comes from the still harder iron. Though a hard track might bruise the heels, it is evident the broader surface would afford some relief. Tips leave the natural guards intact, without the evils which follow pressure on the heels. It is evident that the prolongation of so light and narrow a piece of iron as a racing-plate cannot afford any protection. Some of them weigh as little as one ounce, and those weighing four ounces are very heavy. This thin band of metal is somewhat analagous to a foreign substance between the heel and shoe in the human foot; and we all are aware of the inconvenience a very small thing will be in such a case.

The race-horse strides from 18 to 26 feet, and the fastest in a "brush" go at the rate of a mile inside of one hundred seconds. The weight of horse and rider will be over one thousand pounds, two-thirds of this being borne by the fore legs. The velocity and the weight combined make the jar tremendous, and frequently the heel of the plate is torn from its fastenings and injures the leg. Several race-horses have been ruined in this manner—the most notable, perhaps, the great Longfellow. The use of the racing-tip obviates all danger of this kind, and consequently the advantage, in this particular, is worthy of consideration. A few horses have been able to run faster in their training-shoes than when plated—doubtless arising from the broader surface of the shoes distributing the jar over a greater width, and those horses having a light wall.

Dan Mace, in his letters in the New York *Spirit of the Times*, ascribed Fullerton working badly on one occasion to the heels of the shoe having sprung so as to give a wrong bearing. This horse has a light wall, and, as it is known that white feet are not usually as strong as those of a darker color, a trifling displacement of the shoe would affect him, when another horse with a stronger foot would be able to bear it with impunity. It is hardly necessary to present further arguments on the advantages of leaving the quarters completely unfettered, and retaining the whole capacity of the heel to obviate the bad effects of concussion. It will be as well to consider the objections to tips being used on horses which are required to work fast, and, perhaps, there may be something so fatal to their general use that it may be necessary to adopt the old shoe, or a remodification of it. Stonehenge's objection, that the frog and heel would be worn so thin as to lame the animal, does not hold good so far as I have experimented. It is evident that a macadamized, gravel-covered road would present most friction, and as some horses slide their feet along in making the stride, the wear would be very severe. The subject I have tried was particularly prone to this manner of progression, and an ordinary shoe was soon worn out on the Oakland streets, which are entirely paved with macadam, and some of them covered with sharp beach gravel. I tested the "elastic shoe" on him, a cast-iron shape, with a groove in which was filled a composition of rubber and wool. In less than a week the elastic material was entirely worn away, and the greater portion of the soft iron at the toe. The animal was three years old in 1875, thoroughbred, and had a good deal of trotting action, being able to trot a mile in four minutes, showing at times a much faster gait on the road. He presented a good opportunity to experiment with, and I tried several varieties of shoes, preparing his feet and nailing them on myself.

I began with the shoe first described, viz: two plates of cast brass or copper, with a stratum of rubber between. In the Fall of 1875 I commenced the use of tips, driving him frequently on the road. The first used were cast something of the form shown in the cut in the next chapter, but without the depression for the introduction of the rubber between the sole and the metal. The tip extended back a short distance behind the point of the frog, and the foot was prepared by cutting the horn down where the tip came, until the ground surface of that and the horn at the heel was on the same level. It being slightly wedge-shaped, most of the horn was cut away from the toe, where it could be cut down more safely than at the posterior portion of the tip. Shod in this way there was not as much wear at the heel as the natural growth, while the toe of the tip would soon be worn to a feather-edge. All winter he was thus driven, doing all the "running around," to town, and generally on Sundays long drives on the gravelled roads. This was varied by an occasional brush on the track when the race-horses in training needed company, and several times he was taken across the bay and driven over the cobble-stones, Belgian pavement, red-rock roads of the Park, and the Cliff House road to the Bay District Course. As a further experiment, I nailed tips cut from a slab of copper three-sixteenths of an inch thick, and covering the whole of the foot from a short distance back of the frog, there being a small, triangular space for the apex of the frog to come through. He could not trot as fast in this as the heavier tip, but could run faster. The copper sunk down in the middle until it touched the sole, but as it was quite concave, there did not appear to be any ill effects following, as the main pressure was thrown upon the edges. Like the others, the wear came upon the front portion, and it remained upon the feet until that part and the horn at the toe were worn to quite an acute angle. Some writers who have treated of the subject of shoeing, have recommended the new shoe being formed in this manner, claiming that the wear was due to the horse dragging his toe along the ground, and that rounding the shoe to the shape of the one that had been worn, enabled the horse to travel more easily, and did away with the tendency to stumble. The worn tip made from the tough material proved that this view was incorrect, as the edge was turned up so as to lap partially over the horn, showing that it was the turning of the foot from the heel to the toe which did it, and also proved the importance of having an edge which would not slip as the foot left the ground. There are horses which dwell in their action, which will require the beveling of the shoe, and this we will take into consideration hereafter.

CHAPTER III.

Different Forms of Tips—Effect of Weight on the Feet—Miles' Essay—Etc.

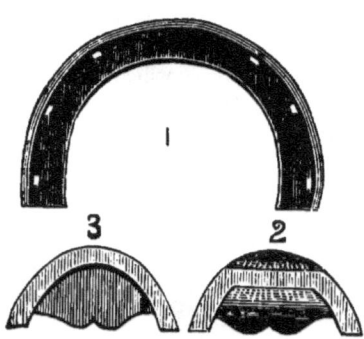

The cut represents two tips, No. 1 being such as is preferable for the race-horse, and some trotters, when in training; 2 and 3 the ground and foot surface of one for trotting colts which have such a gait as requires this shape. The first representation is rather too long for the race-horse, and in my practice I have found it better not to have the tips come further back than the point of the frog, and four nails, all that are required to properly fasten it to the foot. It is made light and with a swedged rim to give a better hold of the ground, with only projection enough to effect this, as a deeper rim would throw too much weight on the wall. The only thing required in the race-horse is to protect the toe from wear, and the lighter the tip, when sufficiently heavy to stand the strain, the better it is. For seventeen years I have followed the plan of nailing from the inside of the wall, and in HORSE PORTRAITURE, and in articles written prior to the publication of that work, have recommended that system, and given my reasons for the preference. At that time I thought it original with me, but have since found that it was the French manner of nailing, and for centuries back the Arabs fastened the shoes on their horses in that way. The form of the nail which the Arabs use compels the nailing through the walls, as the shape of it renders it impossible to drive it so close to the edge as our blacksmiths do; it is more like the old-fashioned clinch-nail, which the country blacksmith of forty years ago made by hand, and the wall is perforated by it at an acute angle.

Dr. Mayhew, an English veterinarian, has shown the advantages of this plan, and his advice was written about the same time I adopted it, but of which I was ignorant until a few years ago. It does not require long argument to prove the advantages of this system of nailing. The wall of the horse's foot being formed of thin layers of horn, agglutinated together, a wedge-shaped piece of iron, which the horse-nail is, when forced with the "grain" has a tendency to split them apart; and how often do we see the whole side of the horse's hoof torn off and clinging to the shoe?

Driven from the inner side through the layers, the wall is perforated and the clinch forms a rivet, which makes the foot actually stronger.

The liability to "prick" the foot is very great when the nail is started from the edge and has to be driven on a curve until it "gets a high hold," and if the sensitive tissues are not wounded, oftentimes it comes so near that the pressure makes the animal lame.

A lower hold, with the whole strength of the wall, is far more effective, and since following the practice my horses have retained their shoes much better. With the tip and four nails I have never had one come off until it was worn entirely through. While figure 1 represents too long a tip for the race-horse, it is the right length for a trotter which requires a toe-weight, as in that case the space between the second and third nail is required to hold the strap which secures the weight behind. A slot is filed in the metal, through which the strap is drawn, and the hook of the "Eureka" toe-weight, at the toe, makes a firm junction. This will be fully considered when I write of the uses and effects of these lately-invented trotting appendages. Figures 2 and 3 represent a peculiar tip, and while one of the shape of figure 1 with the toe-weight attachment will be found, in a great majority of instances, all that is required for trotting colts and older horses, many young animals will be much benefited by wearing such as these cuts represent. Very frequently colts of great promise to make fast trotters have so little knee-action that they point and dwell in their stride, and after months, or perhaps years, of careful education, still retain so much of this faulty action as to greatly interfere with an increase of speed. Rattles, strings of small bells, weights, and all the appliances, fail to remedy the defect, and after patient endeavor the trainer is forced to give them up. Road-driving sometimes overcomes the defect, but in a majority of instances the habit becomes fixed, and if the animal is strongly urged is very liable to get to hitching, single-footing or shuffling along in the attempt to go faster than his gait will permit. Horses of this kind of action are very liable to cut themselves on the coronet, or bring the hind foot in contact with the shoe on the fore foot. This comes from the dwelling habit, the front foot not being picked up fast enough, the hind foot catching it before it is raised.

Figure 2 represents the ground surface of the tip, with the beveled toe, and extra posterior projection to give weight. The beveled toe is to quicken the stroke, the weight to induce higher action. Several years ago, when living in Iowa, on the banks of the Mississippi river, Mr. Robert Bonner sent me a pattern of the "rolling-motion shoe," with calkings so that it could be used on the ice. There were four calks on the shoe, those in front being set back fully as far as the transverse bar shown in the cut. It astonished me with the effect it had in quickening the motion of the front feet. On the hard, icy surface this effect would be more marked than on the softer track or road. With an ordinary shoe, horses will "pick up quicker" on the ice, and with this the effect was to subject the driver to a perfect shower of small pieces of ice, thrown with a velocity which made the seat of the skeleton sleigh an "anxious-seat" for the reinsman. A sulky had to be substituted, in order to obtain a higher seat, to avoid the peppering, and a stream of glittering particles would follow the wheels. The cause of this is apparent. With the calkin on the toe, it lengthened the point over which the foot had to roll, and placing the fulcrum further back, gained a shorter and consequently a faster stroke. In that part of Iowa were many Germans, and one of them had acquired a wide reputation for the wooden shoes he fabricated. They were much easier to walk in than those made after the usual pattern, and the whole secret was his placing a bar across the bottom of the shoe, below the ball of the foot. It was analogous to the set-back calkins of the rolling-motion shoe. The rigid material did not permit the bending of the foot in the manner the leather sole does, and in lieu of placing the fulcrum at the extreme point, it was back some three inches, requiring much less force to raise the heel. The Chinese shoe is a further exemplification of the principle that an unyielding sole is made easier for the wearer to travel upon when the toe is beveled like the tip.

I found that a tip made after the plan shown in the cut, but resting on the sole, made the horse cripple after it had been worn for a time; and one in which the metal was cut away had the drawback of permitting the wet clay to become impacted between it and the sole, so as to produce the same difficulty. Figure 3 represents the foot surface, the outer rim half an inch wide, and raised above the portion which covers the sole an eighth of an inch. In this depression is fitted a piece of the best India-rubber, a little thicker than the depression, and when the tip is nailed on it keeps its place perfectly, preventing the gathering of dirt "between the sole and the metal," and also giving an elastic material which does not injure the sole with an injurious pressure. The toe of the ground surface, being beveled, has another advantage, as it is not so likely to wound the coronet, or the lower part of the ankle; neither is it so apt to strike the horn of the hind foot. Closely watch the action of the fast trotter—the best place to see, when seated in a skeleton wagon—and you will observe that while the fore foot is raised from the ground, the hind foot on the same side is thrust under it, some horses going outside and some inside of the center of the front foot. When the knee is fully bent, the fore foot is raised higher, and the hind foot strikes the ground much in advance, and at nearly the same time as the opposite fore foot touches the earth—so near that it is difficult for the eye or ear to distinguish any difference. It is about the middle of the stride when the feet come together, and further along the foot is raised until, in some horses, the shin is struck, not unfrequently as high up as the hock. When the shin is struck, it is generally done with the lower, outside edge of the shoe, and there is less danger of this injury when tips are worn. Any intelligent, close-observing trainer of trotters will have noticed how colts endeavor to avoid the injury. Some will twist themselves sideways and trot like a dog, one hind foot going inside of the front ones, the other to the outside: others make a sort of a jump behind, both being detrimental to speed.

Boots are applied, and there is little doubt that the improvement in the trotters of the present day is greatly owing to the more intelligent use of these adjuncts. Still, it is manifest that boots on the hoofs of the hind feet, extending above the coronet, on the pasterns, ankles and shins must, more or less, hamper the animal wearing them, and if the difficulty can be obviated by a change of shoeing, it will be a superior method of overcoming it. But if this change in the shoeing gives a wrong bearing, an unnatural set of the feet or limbs, the remedy would eventually be worse than the disease. The use of tips present a better opportunity to modulate the action than is possible to accomplish with shoes without endangering the feet and limbs. An illustration, and one which has struck me with the greatest force, is the change in the action of the colt when first shod. He has been broken and driven some, before anything is placed upon his feet, and his trainer will tell you that there will be a favorable change whenever he has the iron fastened to his hoofs. In ninety-nine cases in a hundred, the result will be as predicted, and the shoes, weighing in the neighborhood of a pound each, will increase his speed by several seconds. I have found a tip of not more than six ounces to have the same effect. Again, a tendency to pace is overcome by a heavier shoe or the resort to something else which has been found to have an analagous effect.

One of the most successful trainers I ever knew in converting the pacer to a fast trotter, informed me that to run the horse with feet weighted, until he became too tired either to run or pace, was the most effectual method he had ever found to overcome the propensity for the lateral manner of progression. This proves that a heavy shoe or heavy toe-weight is inimical to speed, either running or pacing, but is adapted to the trotting gait, and the horse, finding he can get along easier when thus encumbered, naturally tends to relieve himself by adopting the action suitable to the changed condition. and that which tired beyond endurance in the other paces can be sustained at the trot.

This is also further proven by the other methods which trainers employ to change the pace into the trot; the old plan was to strew the road with rails, and ride the animal over it; another, to practice the horse through loose sand or deep snow; and lately, in Texas, a very fast trotter was converted by driving him on the beach when the water reached his knees. The latter method is evidently a very effectual one to cause the horse to bend his knees, and the theory of the effect of weight on the action and the practice coincide. It is manifest that the knees must be bent more to enable the horse to get through the water easily, for if the leg was pushed along, the resistance of the fluid would be great; consequently, the horse soon learns to pick his foot up as nearly perpendicular as he can, and thrust it well forward. The most approved theory is that the weight influences the action the most strongly where the heaviest weight is placed, and with shoes made much heavier on the inner quarter, the striking the knee will be more likely to follow, and a horse which hits his knee with an equal shoe, will avoid it when the outside is made the heaviest. It will necessarily follow the adoption of this hypothesis that weight on the toe will have a greater influence on the action than the same amount distributed over the whole foot, and though the present form of the weight was invented to obviate the bruising of the heels, from the older-fashioned kind, it. was based on the scientific principle of the corelation of forces. Thus, a bullet with one hemisphere cast of a denser material than the other, will fly in a curve, the shorter radius being on the light side. The lighter the side the greater will be the effect, and if the power could be applied so as to overcome the attraction of gravitation, such a ball would describe a horizontal circle.

As has already been shown, while weights on the feet increase the speed, and establish the propensity to trot in preference to run or pace, there is danger to the legs attending the use of it, and many promising horses have been irreparably injured. The better acquaintance with the weights has lessened the danger; when first invented it was held necessary to have a heavy shoe, to which they were added, and now it is found that a lighter shoe does away with the necessity of so much weight, and the same result follows. More than twenty years ago I had a pair of weight-boots, which were made in Boston. They were simply a quarter-boot, filled with lead; they were actually a detriment, from the weight being mostly on the wrong portion of the foot, and after a horse wore them a few times he would be very sore from the pounding on his quarters. This should have been foreseen by any one who gave the subject much thought, as it increased the concussion, and, between the weight and the heavy, thick-heeled shoe, the blow came from both above and below.

That the present toe-weight does effect the purpose, while the former was ineffectual, establishes that the location is all-important; and, furthermore, that it does not depend so much on the amount of the weight as the balance being properly sustained. If a twelve-ounce shoe and an eight-ounce toe-weight has the same effect as a two-pound shoe and a pound-and-a-half weight, it is obvious that a tip which only weighs six ounces will give the same equilibrium to a four-ounce weight. The whole weight of the tip comes in the right place, and adds to in place of lessening the advantages.

The trouble was to fasten a toe-weight to a tip, and until I became acquainted with the "Eureka" fastening I gave the whole of my study to combine the weight in the tip by making it still heavier than the cut shows. There were objections, which I will explain hereafter. Before touching on this I will give some further illustrations of the benefits arising from leaving the posterior portion of the foot in its natural state.

Mr. Miles, in his "Treatise on Horse-shoeing," writes:

"I know that many smiths are fond of what are called 'open-heeled shoes,' which means shoes with straight heels, wide apart, and projecting beyond the hoof, both behind and at the sides, and the only reason I have ever heard in favor of such shoes is a very bad one, viz.: that the horse requires more support at the heels than he gets from the hoof. But you may depend upon it that nature has made no mistake about it; and if the horse really wanted more support than he gets from the heels of the hoof, he would have had it. But I think I shall prove to you that this kind of a shoe, instead of being a benefit to the horse, is a positive evil to him; it interferes with his action, and exposes his sole and frog to serious injury from stones in the road, and the projecting portions of the shoe become ledges for stiff ground to cling to and pull the shoe off. More shoes are lost through this mischievous projecting at the heels than from all other causes put together.

"Let us see how it is that these projecting heels interfere with the horse's action. It is not necessary for this purpose to trouble you with the anatomy of the foot, but merely to state that all its parts are joined to each other in such a manner as to form one great spring, and that the foot is joined to the leg by the pastern and coronet bones, in a direction slanting forward, which brings the foot a little in advance of the leg, and places the heels in front of a line dropped from the centre of the fetlock joint to the ground :

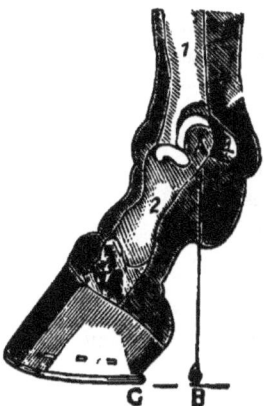

"Figure 1. The shank or cannon bone. 2. The pastern bone. 3. The coronet bone. 4. The sessamoid bone. A. The point where the weight of the horse would fall upon the upper end of the pastern bone. B. The point where a line dropped from A would meet the ground. C. The heel of the hoof. Now, it is clear that the weight of the horse will fall upon the upper end of this slanting pastern bone at every step; and the bone, having a joint at each end of it, will sink to the weight thus thrown upon it, and break the shock both to the leg and foot; but if the heels of the shoe are longer than the heels of the hoof, the projecting pieces of iron will meet the ground further back than the natural heels would have done, and will check the sinking of the pastern bone, just as an upright pastern does, by bringing the heels too much under the center of the weight, which causes the horse to stop short and go stumpy. If you wish to avoid these evils and keep the horse's shoes on his feet, you must bring in the heels, and let the shoe strictly follow the form of the foot, whatever that form may be."

This argument of Mr. Miles is a forcible illustration of the bad effect of doing away with the natural bearing of the horse's foot. and, strongly as it favors the accurate fitting of the shoe to the foot, is a still better exemplification of the necessity for leaving that part of the fore foot as nature made it, unhampered and free to perform its natural functions. The illustration shows the position of the foot and leg when the animal is standing; when in motion, and particularly when going fast, there is far more necessity to guard against the evils of a wrong bearing. Then the foot is thrown forward and strikes the ground at the farthest point it will reach. The formation of the fore limbs from the scapula down is such as to break the jar as much as possible. Unlike the hind legs, they are not rigidly attached to the bony framework of the trunk, but are bound to it with elastic bands of muscles.

The angle which the humerus and scapula form is the same as the elliptic steel spring under a carriage; from the elbow to the ankle the bones are curved, and the sloping pastern, bending down when the weight is thrown upon it, complete the protection above the foot. Were the foot a solid, unyielding body, all the protection alluded to would be rendered useless. It would then be like striking with a hammer which had a springy handle, the blow being more acute, more stinging from the spring. But the most admirable contrivance of all to obviate the ill effects of the jar, arising from the rapid concussion, is in the foot in its natural state. The opening paper of this essay illustrated this, and the first cut explains how the spring of the foot aids the elasticity of the frog in counteracting the difficulty. At the risk of repetition it will be well to give this matter the fullest consideration, as, in my opinion, this is the most important matter to thoroughly study. All of the best veterinarians, and a great majority of the writers on the pathology of the horse's foot, have recognized the importance of retaining this safeguard, and have recommended various plans of shoeing to obviate the difficulty of a rigid band when nature intended there should be the fullest motion.

Mr. Miles, and many others, advise few nails being used, and those, as much as possible, driven near the toe; others, a shoe with a joint at the toe, and another patented a shoe which was held in its place by screws which clasped it to the outside of the wall.

All of these have been guided by the belief that the posterior portion of the foot must be protected by an iron rim, and though the quotation from J. H. Walsh, heretofore given, proves that he had a faint idea that this protection might be dispensed with, he had not faith enough in it to give it a trial himself, though he signified his wish to see it brought more fully into use. By returning to the matter quoted from "Stonehenge," it will be seen that in some instances he knew that shoeing with tips was successful, while he does not mention one where there was a failure. Miles is afraid of the horse injuring the sole and frog with a narrow shoe, from the facility with which small stones could be introduced and bruise the exposed parts. Mr. Miles would have seen ten times the danger in a tip, but in his instructions to the smith he directs the paring away of the sole and the cutting away the bars until they are on a level with the excised sole. Left with the protection which nature has given them, there would be little danger of the kind, and though a sharp-pointed flint or stone might do injury, a still sharper one would injure, no matter how wide the web of the shoe was made.

CHAPTER IV.

Cure of a Sprung Tendon—Results of Experiments—Etc.

The horses which have most of their work on the tracks, wearing the tips, would not prove the fitness of them for road use, and in order to test them on as unfavorable ground, and under as untoward circumstances, I tried them on three three-year-old trotting colts, and the same thoroughbred spoken of previously. The latter was the most effectual subject for experiment, as he was driven long trips, and his action naturally was so low and sliding that he soon wore an ordinary shoe completely out. In coming from Chicago the horses were only taken off at Omaha, and never left the car until they arrived at Oakland, those who had them in charge not following my instructions to make the stops between. There were eleven horses in the car, and the long journey, and being continually on his feet, caused this colt to go over on his knees. He was a sickly colt, and the summer before, the large green-head fly, which is such a pest in the neighborhood of Chicago, very nearly killed him. Though two years old at the time I brought him to California, he had the appearance of a yearling, and this physical weakness doubtless was the cause of the trouble, as the other horses showed little of the ill-effects of the long journey. While he wore shoes this difficulty appeared to be increased, and he had the appearance of some old stage-horse which had been sent along for years over the hard roads. Shortly after the use of tips on his fore feet his knees began to straighten, and now they are only a trifle out of the perpendicular. During the summer of 1876 I sent him to Sacramento, where another of my horses was in training, and the man who had them in charge, like a majority of others, was obstinate in sticking to the old system. It was not long until he "sprung" one of the tendons between the ankle and foot, and he had to be thrown out of training. The shoe had evidently been put on without proper care having been taken in giving a true bearing, the greater pressure being on the inside quarter and heel, and the horn, on that side, was forced up until there was quite an elevation at the coronet. When I got him home the tendon was considerably enlarged, but the inflammation had subsided. I treated the enlargement with the biniodide of mercury preparation, and one application reduced the enlargement so that it could not be detected that there had been an injury. I poulticed the coronet with boiled turnips, and, until I commenced to drive him again, allowed him to run barefoot in a small enclosure. The foot, below the injury, showed the effects of the wrong set of the shoe, and, the result of the inflammatory action, the frog was shrunken and hard. Before turning him out I pared the foot down at the toe, and shortened the wall at the heel until the frog and bars touched the ground first.

A few weeks of this treatment had a surprising effect. The hard elevation at the coronet became soft, and ere long it returned to its normal condition; the frog resumed its elasticity, and there was a perceptible widening of the heel above. I shod him with a tip which would weigh six ounces, and fixed so as to attach the Eureka toe-weight, my object being to give a further test of track and road work on a trot, before he was put in training to run again. Like a majority of horses, he wore the hind shoes much faster than those in front, the outside of both being most worn. The usual shoe was put on his hind feet, with slots to fasten a toe-boot on. His work was varied, partly on the track, at which time he was hitched to a sulky, and when on the road a wagon with two persons in it was the vehicle used. The streets of Oakland are noted for the rapid attrition of the iron, but even daily driving of this horse on them did not wear away the horn as fast as it was produced. This result on the fore feet did not lead me to the belief that tips would answer on the hind, there being so much difference in the action. The front feet strike the ground nearly perpendicular to the plane of the foot, while the hind feet are slid along, especially when horses go "close to the ground." But, determined to make the test as thorough as possible, when the hind shoes were worn entirely through, I substituted tips, expecting that they would not do. These were set on the 13th of January, and the 4th of February they were worn entirely through, without the heels being affected. They were wider, and not so thick as the tips in front, and the nails were left in the horn just as they were driven, minus the heads. At the date of writing this it has just been a month since the heels of the hind feet were exposed, and there has not been the slightest soreness, though scarcely a day has elapsed that he has not been used on the road.

Two of the trotting colts—fillies—had been shod the Fall previous with ordinary shoes; the other never wore anything but tips while I had him. The fillies were jogged a short time barefooted, and when the tips were put on they were prepared so as to fasten the Eureka toe-weights on them. Both improved very rapidly before the weights were applied, and with a six-ounce weight there was an increase of speed. One of them was by a son of Mambrino Chief, out of a Blackbird mare; the other by a Blackbird horse, her dam a thoroughbred mare. The former was a rapid, short-gaited filly; the other the reverse, going with her head low and a long, sweeping stride. The Mambrino I tried with a nine-ounce weight on each fore foot, and she trotted three quarters of a mile in 2:04—a rate of 2:45 to the mile. The year before she lamed herself in the hind leg in the stall, and at times she would show a little of the same trouble, and after this fast drive she favored her leg so much that I threw her up, and she was turned out to run through the Winter. The other filly also improved rapidly, but never showed such speed, being able to trot the mile in about 2:50. Both of these were shod with the ordinary shoe behind. The third one was a colt by the same Blackbird horse that got the filly, his dam a grand-daughter of Rysdyk's Hambletonian. He came from Chicago the Fall before with a coat like the winter raiment of a buffalo, and was a little "chunk of a colt" which evidently had not been favored with much to eat save the sour prairie grass which is the main feed in these pastures. He was full of worms and sickly during the Spring, but grew fast, though it appeared impossible to get any flesh on him. Like his half-sister, he was a long strider, and his improvement was also rapid. At times there was an inclination to single-foot, and at first he had a dwelling motion in his stride. The application of the six-ounce toe-weight would correct the first, and a tip beveled at the toe as shown in the cut remedied the second. Being so poor, though he ate plenty of grain, I restricted his fast work to half miles, and in a few weeks handling he dropped from a "three and a half gait" to half a mile in 1:21. I found it much easier to control his gait with the tips than I ever had with shoes under the same circumstances. In all of these colts' feet the frog was full and elastic, and the hoofs retained their proper shape. This colt won a three-year-old stake under many adverse circumstances.

The case of Hock-Hocking presents the value of tips in a very clear light, and though he received irreparable injuries before wearing them, the history will show that he could be trained and run with them when, with ordinary shoes, he soon became lame. When I first got him—the Spring he was four years old—his feet were out of shape from the irregular growth of horn, and owing to his antipathy to have his feet handled the heels were too high and the quarters somewhat contracted. Then it was a difficult job to shoe him, and even with a twitch on his nose he gave the blacksmith a good deal of trouble. His work was on a hard track, much harder than it appeared, and he split his foot from the coronet to the ground. It is needless to recapitulate the races he ran. He was continually shattering his hoofs, and, notwithstanding the support of copper plates screwed to the wall, they would fracture with almost every gallop. He started in a race of two mile heats with Waterford and Woodbury, was close up in the first heat in 3:36¾ literally on three legs. In the second he split his other fore foot, and started the small metacarpal bone from the larger.

The Spring of 1875 Mr. Dunbar operated on him, and for months, following instructions, we soaked his feet three times a day in hot water, walking him on the beach when the tide was in, and taking every pains to grow his feet anew. I put him in training in the Fall, with shoes of the usual weight used in training race-horses, but after working a short time he went lame. I gave up all thought of running him in the four-mile race, but when it was postponed from Christmas and New Year's until the 22d of February, and then having adopted the tips, I concluded to try them on him. His ankle was very much enlarged from the injury to the splent bone, the lower point of which could be felt between the middle and back tendon. In this training his lameness never appeared, and though worked very hard, he started in the race in fine condition. I am not alone in the opinion that if he had been fairly dealt with he would have won the race. He exhibited scarcely any distress after the first heat, and was only three seconds behind the winner. After that, when galloping on the road during a protracted rain, he struck a stone and wrenched his ankle, which made a permanent enlargement of the joint. To test the tips in a case of this kind, he was again put in training, and up to this date is doing well. Since wearing the tips there has been nothing like a quarter crack, and I am well satisfied that it is impossible to spring one of these troublesome affairs when the tip is worn; and, furthermore, I have never yet had a race-horse which "broke down" or had a "bowed" tendon unless there was something wrong in the feet.

CHAPTER V.

From Shoes to Tips—Further Satisfactory Tests.

I am not prepared to say that one set of experiments, conducted by one person, and that person liable to an undue prejudice in favor of a new departure, is conclusive testimony of the value of tips, under all circumstances. To sustain the theories, however, I have given the results in cases which have come under my practice, and conclusive as they are at present to me, there may something arise to show that there are defects which will eventually prove objectionable.

The whole system is so simple that apparently it would not require long time to either demonstrate the superiority of this method of shoeing or to establish some striking defect. But, as many are aware, the results of a few experiments are oftentimes delusive, and only after the severest scrutiny can an authoritative opinion be reached. This has been the reason that I have delayed these articles, wishing to give the matter more study, and to practice still further with the horses. I have been extremely anxious not to "jump at conclusions," and determined to test, as fully as in my power, the practical working, and noting the effects of changes from tips to shoes, and from shoes to tips, endeavor to arrive at a proper estimate of the relative merits of each.

The more I studied the matter, the less reason I saw for the theory proving erroneous so far as the well-doing of the feet are concerned, but I am well aware that for road-horses and trotters there might be another drawback which will be fatal to the use of tips. This is the effect on the action, and there may be a necessity for weight on the posterior portion of the hoof to enable a horse to trot fast, and the absence of the iron at the heel might induce a greater tendency to shuffle or single-foot. Every trainer who has had much experience with colts, knows the care which is requisite to keep them trotting square, and how frequently, notwithstanding his efforts, they will "get off their gait," and retrograde in speed. They will do so when it will puzzle the most acute to give a satisfactory explanation, or detect the cause of the change in the action.

So far as my experiments have shown with the trotting colts, there was no greater tendency to those hindrances to speed with tips than with shoes, and those which I have instituted since writing the former articles have been with more of a desire to test this than anything else. The result of my own trials were the only ones I had for data, as there is such a violent prejudice with *horsemen* against innovations, particularly when that change is an entirely different system of shoeing, that the most enterprising were loth even to listen to arguments in favor of the change.

It is an easy matter to convince any person of ordinary intelligence that the feet are benefited, when a comparison is instituted with those which wear shoes, as that is too palpable to be contradicted. Especially is the variation from the natural foot, which shoeing causes, more notable when the animal has worn the ordinary shoe a sufficient time to change the entire structure of the foot, for I am inclined to the belief that after a horse has been shod for a few years the foot can never be brought back to its original form. The injuries may be palliated, but never entirely overcome.

But while this advantage to the foot is susceptible of demonstration, the bearing it has on other subjects of importance is not so evident, and people are prone to ascribe difficulties which have arisen from other causes to the new departure. I have been more anxious to convince myself than others, and have watched as carefully, or more so, for radical defects than for arguments to sustain it. In such a case I would have been ready to give them publicity, as I have no pride in sustaining a position once taken, if that position has not a logical foundation. Mere assertions, however, will not force me to surrender convictions, and the *ipse dixit* of the most celebrated professors has little weight if not sustained by proof of the soundness of the objections. If I find that the horses I have tried are benefited by the change from shoes to tips, it is fair to infer that others in like circumstances will also find a corresponding advantage; and though to prove the absolute improvement, all kinds will have to be tested, the chances are favorable that the good results to the few augur well for the success of the plan to the many.

Notwithstanding the efforts of breeders to intensify the fast trotting gait, and in which they have succeeded beyond the faith of the most sanguine, it is, nevertheless, in very fast trotting easily subject to changes which are inimical to improvement. There is a tendency, as all trainers know, to forsake the true trotting step, and "singlefoot" and "hitch" and "scramble," and perhaps one or more of a dozen things which prevent further improvement. The animal may retain its square action, and yet step so short that it cannot go fast, regardless of how rapid the stroke may be; or it may stride so long as to "dwell," and in this case there is usually a deficiency of "knee-action."

Intelligent trainers are aware that all of these things may be partially remedied, in many cases entirely overcome, by a change in the shoeing—still more by the wearing of weights and boots; and yet there are animals which persist in the faulty action, despite of all the appliances of ancient or modern days. I have faith that more can be accomplished with different varieties of tips, especially when colts are the pupils, but there may be individuals which will demand the application of the full shoe to counteract the faulty action which the animal persists in, though used with many different styles of tips.

The necessity for a full shoe for trotting horses which have a shortness of stride, or a tendency to forsake the true action may arise, although thus far I have not found it so in my practice, though that has been confined to a few animals, and those of a kind which I would consider the most likely to be benefited by the greater weight in the shoe. The horse I have experimented the most with is X. X., the colt alluded to in previous chapters as having been used on the roads with tips, in order to test them where there would be the maximum of wear. Since writing the former articles, I have used him with various tips, and changed him to shoes in order to see if the additional weight would not influence his trotting action more favorably. Desiring that Lady Amanda should have company when galloping last spring, I gave him a rough preparation, and ran him against Emma Skaggs and John Funk, a race of mile heats. This he won in 1:50—1:50½, and could have run something faster. When in training I commenced with tips, and as they wore out let him gallop barefoot, and he ran in the race without anything on his feet. The tendon which was sprung got a little "hot," and he was slightly lame in that leg, but afterwards we galloped him in a sulky when the heat and tenderness diminished, and with the resumption of road work it got apparently entirely sound. He has never been really well, and though a package of Professor Going's worm powders brought away a great many of these troublesome parasites, there appeared to be a fresh recruit ready to take their place. His appetite was voracious, and though fed very high, when his grain and hay were consumed he would attack his bedding, the soiled portion being devoured with as much avidity as the best Oregon oats, or the brightest provender. When at Palo Alta, not long ago, Governor Stanford informed me that his colts were ridden of worms by using lime-water, and X. X. has, since his feed has been wetted with it, done much better.

Still he did not improve in his trotting, and when urged to go faster would hobble behind, and mix his gait. When galloping him, harnessed to a sulky, he was shod with very light steel tips, scarcely heavier than plates, on both fore and hind feet, and he was driven with them on the road. These I replaced with a heavier set, but he still persisted in hobbling. I then had him shod with the ordinary shoe, weighing 18 oz., and still he hobbled when driven beyond about a four-minute gait. I applied toe-weights of from six to twelve ounces, without improvement, while his knees went over more than they had ever been. He had the least knee-action of any horse I ever saw, almost dragging his toe along the ground, with a sort of twist when he took his foot up, turning it to the inside as though he did it to avoid contact with the road. I kept the shoes on longer than I should, in order to give him time to become accustomed to the heavier weight, but being convinced that it was useless to try them further, I replaced them with tips.

From his heel being lowered so as to prepare the foot for the shoe, the tip could not be set properly, and I was apprehensive that the less amount of horn might cause bruising of the sole. But from the first day he traveled better, going more squarely, though he still had the twist when picking his foot up. The next time I put tips on him I let them extend to within an inch and a half of the heel on the outside quarter, and on the inner shortened them so that the toe was only covered. There was scarcely any of the twisting motion left—none at all on the right foot, and he has shown a decided improvement, in every respect, since he wore those one-sided tips.

I have also been trying the difference between shoes and tips on the Alhambra filly previously mentioned. When she became lame she was taken to Mendocino County and turned out, running there through the winter and until about the first of July. After jogging her for several weeks, I moved her through the stretch and found that her stride was shortened from what it formerly had been, and the only explanation I could give was, that she had acquired the habit from favoring her leg while lame. She was wearing tips with holes in them to apply the Eureka toe-weight, but not increasing her stride to a length that was satisfactory, I concluded to try the full shoe, on her fore feet, weighing about 18 oz. each. The fastest mile she trotted while wearing the shoes was 2:52, and they were retained nearly three months in order to give them a fair trial. Three days after the tips were put on she trotted a mile in 2:48¼, going very steadily and apparently well within her rate. She kept at about the same mark for the next month, though still striding much shorter than she did when a three-year-old; and being desirous to experiment further, on the 10th of December I had a set of front shoes put on weighing 22 oz. each. These are still on her, and I cannot say authoritatively what will be the result, but I am inclined to the belief that they are not going to increase her stride any more than the tips did. She certainly goes much easier in the latter, and the four months wear of the shoes can readily be noticed in the different appearance of her feet, there being an evident contraction of the heels, with the frog much narrower than when the shoes were put on.

I feel very confident that the shortening of the gait was caused by her being turned out in the field (a very large one—two hundred acres) while she was lame, and running when she had to favor the limb. She would run for half an hour at a time, careering over the hills, and after feeding a little would resume the play.

Having entirely recovered from the lameness, I think in time she will resume her former action. I expected in this case to find the shoes preferable to tips, if even they were not so good for the feet; but so far I have been disappointed in the result. When I replace the tips I will extend them across the foot, and thus get as much weight as possible near the toe and leave the posterior part of the foot unhampered.

A horse assuredly puts his foot down more gingerly when the iron comes under the heel, and the only reason I could see for the full shoe extending the foot further than a tip, was the greater amount of weight in the whole mass.

The effect of weight on the feet is a difficult problem to understand, in its bearing on the speed and action of horses, especially trotters, and the only reliable way is to bring it to an actual test with trials on different horses. But one thing few will question, and that is, that a load of iron on the feet is prejudicial to the endurance.

The trainers of race-horses have adopted a much lighter shoe than was formerly in vogue, and some have discarded them entirely. The old-timers not only gave them long, exhausting sweats, under loads of blankets, as *vide* the history of Hanie's Maria, but they wore heavy shoes, their reasoning being that the plates would be a greater contrast and the change give greater speed.

Great as has been the change in the management of race-horses, still greater has been the improvement in those who educate the trotter. The most striking innovation has been the use of the toe-weight, though boots have done a large share in developing the speed.

A person who was without any knowledge of the manner in which trotters were handled, would think that the driver was endeavoring to make them slower, by fastening leaden weights on their feet, and hampering their limbs with so much kersey and leather.

I intend devoting several chapters to the consideration of these adjuncts, and will endeavor to show that faulty shoeing has necessitated the use of boots, when they might be avoided, and that the principle on which the toe-weight depends for its efficacy is in accordance with and harmonizes with light weight better than heavy. The subject is of importance, and is well worthy of much thought.

CHAPTER VI.

Mistakes of Blacksmiths and Grooms—Hard Roads and the Consequences.

So far this essay has been written at wide intervals, delaying the composition in order to try further experiments—the great object being to have the theory corroborated by practice. My opportunity for practice has been limited to a few horses, but, fortunately, these were so different in their action that the paucity of numbers was not such a drawback as it might have been.

The opening chapter was published in the California *Spirit of the Times*, April 29th, 1876. During the last Summer the papers of Great Britain have contained a number of articles on the abolition of shoes on horses, the parties advocating the measure showing the benefit arising from the feet being untrammelled with iron. Their opponents conceded this, but claimed the necessity for protecting the feet when the animals had to work over hard roads or pavements. All which I have seen on the subject has fortified the position I have taken, as the plan recommended in these papers gives the same freedom, while it protects the only portion of the foot which requires artificial protection. In this country there is another element which enters into the calculation, and the question of the effect on the action is one of so much importance that the soundness of the feet in one class of horses is a secondary consideration. The fast trotter is peculiarly American, and in England little is known of his requirements. Here the track-horse represents an immense capital, and yet the value of these, which are kept for the sole purpose of trotting in races, falls far short of that which the road-horses of the country represent. Speed and endurance give the worth to the first; speed is the most valuable quality in the second, provided it is not accompanied by habits which render the speed useless, or by such unsoundness as unfits the animal for rapid driving. So long as there is not absolute lameness, a majority of the owners of fast road-horses are satisfied. A large number of them have so little knowledge of the horse's foot that they cannot tell whether it is in proper shape or not, and trust entirely to the blacksmith and groom for the treatment. Unfortunately, a large proportion of this class of men are not only ignorant of the pathology and functions of the foot, but an immense percentage of them are bigoted in the belief of the efficiency of old-time practices, and too stubborn in this belief to surrender it. They will not listen to argument, and any departures from the old ruts, deeply worn by prejudice, are stigmatized as innovations which are bound to bring disaster.

These men have an influence over owners which is as singular as it is potent. A gentleman may require that his family physician be educated in the latest teachings of the schools, and will only employ one who has forsaken the old-time practices for a more enlightened system of treatment. In any department of business he will give the preference to education, and readily acknowledges the importance of thought and study in every department of life where learning can be brought to bear. But with horses it is different. Ignorance is not considered a bar, and he follows the directions of a man whose only recommendation is that he is acquainted with the stable economy which was practiced a hundred years ago. The feet are mutilated at the forge, the groom stuffs them with filth, or at the best with compositions which soak the horn into a pulp, and what Nature designed to be a firm support, and to have solidity, is weakened by the removal of that which gave strength, and the natural and firm foundation is changed, by erroneous treatment, into a tottering pedestal. The feet injured, the legs give way, and then the poor animal is tortured with virulent blisters; but as these corrosive vesicants compel rest, in that way the animal is benefited, and if his tormentor would let him alone for time to remedy, the cure would be equally as effectual. The hard roads are blamed for the quackery of the groom and the blacksmith, and when the feet and legs are injured beyond recovery, the injury is ascribed to anything but the true source. I am satisfied that the danger of hard roads to the feet and legs of the harness-horse has been greatly magnified, and the great proportion of the injuries which are claimed to be the result of fast driving over ordinarily hard surfaces, are directly traceable to other causes. In this paper I intended to consider the effect of tips and shoes on the action, but, perhaps, it may be as well to devote the rest of the chapter to the elucidation of that portion of the subject, which is so important in road-driving. Hard roads are the rule in the neighborhood of cities, as even moderately-soft highways would soon be cut up by the numerous vehicles which are driven over them. Deep, heavy roads are more likely to injure the trotter than those with a smooth, hard surface, provided the feet have their natural functions preserved, and a continued "speeding" over such ground will result in making the animal slower, if it does not irretrievably injure the legs. A "dirt road," with loose soil enough to fill the concavity of the foot, is recognized to be the best of all for horses to be driven fast upon; but such are rarely found, and, consequently, the driving has to be done over the common macadamized highway, and, in the vicinity of San Francisco, this is covered with broken red trap rock. This, when kept in condition, forms a smooth surface, and when wet gives a firm hold for the foot, though it wears the shoe rapidly. The foundation of broken stone and this covering make a roadway which is nearly as solid as the Belgian pavement, the difference being that the top covering affords a safer rest for the foot, with less liability to slip. During the dry season in California, the soil, unless sandy, becomes as hard as the rock road, and some of the black adobe land is fully as unyielding, drying to the hardness of a hard-burnt brick for several feet below the surface. Unless covered with dried grasses, there is nothing to break the jar, and more trying ground to the feet it would be difficult to find. Notwithstanding this, the vaquero gallops his barefooted horse at a furious rate over it, and goes down declivities at full speed where the English fox-hunter would consider it too dangerous to keep the saddle, and, if forced to the route, would carefully lead his shod steed down the grade. Seeing this, the Eastern horsemen extol the feet and legs of the "mustang," and ascribe the immunity from disease to some natural quality which other breeds do not possess. There was no soaking in foot-tubs, no bandaging of legs. Stopping with cow-dung and clay never entered the thoughts of the owner, and he was as ignorant of "hoof ointments" as he was of all the various shoes which have been invented to "keep the feet in order." His aim was to harden the foot in lieu of softening it, and the sole was filled with tallow and seared with a red-hot iron, when the wear, extending from the toe back, made the animal foot-sore. The legs never gave way, and though the healthy foot had a good deal to do with the tendons remaining unsprung, it would seem as if that were not a sufficient reason for the limbs standing such rough treatment without suffering serious injury.

Doubtless there was another favorable circumstance. The saddle was placed so far back that the weight was thrown on the centre of the body, and to keep it in place the "cinch" was drawn until it was almost buried into the abdomen. The swell of the barrel was where the girth was placed, and the purchase, which the manner of fastening afforded, fixed it in place so firmly that it was immovable. The trainer of race-horses places the saddle on the withers, and the jockey bending forward until his head is almost as far in advance as that of the horse, the whole weight is thrown on the fore legs. This weight at every bound of the horse comes like a blow, and even the hundred-pound boy falls with the force of a battering-ram on the extremities. There is nothing to break the concussion, while the vaquero, though nearly double the weight, is so far back that the yielding spinal column, and the huge, elastic muscles on either side of the back-bone, act as a spring, and the illustration heretofore used, of the spring truck saving the wheels, is again appropriate.

The race-horse trainer says that the horse cannot run so fast with the weight further back than where he places it, though I am not ready to assent to the truth of this until some experiments are tried to prove whether the assertion be sound or not; and until these are instituted I will not argue it. But if the barefooted horse of the vaquero can gallop over such ground without injury to the feet and legs, save the wear the friction at the toe compels, why should the road-horse, with only the weight of the harness, and a part of that of the shafts, give out?

From all the testimony I have been able to obtain, corns and quarter-cracks were never known in horses which were not shod, and in a former chapter the reasons why this should be the case were partially given. A corn arises from a bruise or undue pressure on the part where it is located, and the quarter-crack is the result of concussion; while one part of the foot is bound by the shoe, the elasticity of the coronary ligament forces the fibres apart. The texture of the horn has been injured not only by the condition of the foot being abnormal, from the confinement of the heel, the wasting away of the frog and the removal of the sole, but the soaking, the stuffings, and the application of ointments, have destroyed the life of the deposit, until it has more resemblance to the foot of a dead horse than the tough, elastic material which nature has given. The maceration in the soaking-tub is followed by coating the surface with an oily preparation, and this deceives the eye, as it gives somewhat the appearance of the enamel which nature has provided. This natural enamel is a thin filament of great strength, when compared to the horn beneath it, and while it gives the hoof a degree of hardness to withstand the wear better, it also prevents the moisture from soaking the agglutinizing material which fastens the layers of horn together. The smith has rasped a great part of this glazing away, and the water penetrates readily, is absorbed, the structure weakened, and when the blow comes, the expansion below being hampered, the fibres are torn asunder.

The "road-driver" may say: "What is this to me? My horses are not troubled with quarter-cracks, and I am sure they are free from corns." I have seen more quarter-cracks in California, in four years, than in all my life before, and this probably arises from the long, dry summers, making the customary treatment of the feet more injurious than in the East. That it is not caused by the climate alone is proven by the native unshod horses never having quarter-cracks, and it requires the two causes to account for them being so prevalent.

Corns are far more frequent than owners are aware of, and very few road-horses are free from them. Contraction of the foot, and the shrinkage of the frog, are nearly universal in horses which have worn shoes for any length of time. The narrowing of the heels may be so slight that it escapes observation; the frog may have a fair width, and its functions be nearly destroyed. There is one thing sure: if it is not brought into use, it will not retain its natural vigor, and circumscribing its duties will result in injury to the tendon which lies between the sensitive frog and the navicular bone. In time the internal portion of the foot is affected, and the hard roads are credited with the damage. The injury extends to the ankles, and they become puffed; the whole limb is affected below the knee, and this is regarded as the most complete evidence to establish the fact. The bug-a-boo of a hard road is a convenient scape-goat to carry the bundle which contains the lack of knowledge of the smith and the groom, and the owner has to forego his customary afternoon drive or make fresh purchases for the *road* to destroy.

In previous articles I have called attention to the shoe increasing the concussion on the heel, and I think that any person who will give some thought to the subject will agree with me on that point. With the ordinary shoe the blow comes entirely on the wall, or the junction of the wall and the sole; in the latter case corns result, and in either there is "soreness" ending in established disease of the foot. The bar shoe owes its efficacy to distributing the concussion over a wider surface, and by giving the frog a chance to take a portion of the blow. The seat of the corn is trimmed away until that part of the foot does not bear on the iron, and the relief to the poor animal is immediate. From crippling along with short steps, he strides out with less fear, and the owner thinks that the "round" shoe has some magical properties which enamors him with its use. Suppose that in lieu of the narrow, transverse bar of iron, which only bears on the posterior portion of the frog, the whole surface is made available, and in addition the spring gained by the expansion of the foot is brought into play. It will not require long arguments to prove the greater benefit to be derived from this in breaking the jar, but the objection will be offered that such a course will end in bruising the heels and frog. We have seen that such was not the case in the horse of the vaquero, galloping over hard ground and down hills, in many cases covered with stones, and in my practice with tips, for over two years, it has not occurred.

Where the tip covered there might be a little discoloration, and the sole show a few red streaks, when it had been pared thin in order to shorten the toe, but the heels have remained free from as much as a red spot, and though when I replaced the tips with shoes, this part was cut down fully half an inch, the horn, sole and bars were in perfect condition. There was a change in the substance when compared with the feet of those which had worn shoes, the junction of the wall and bars being as hard as the tip of the horn of an ox, and as tough as it was hard. The foot of the shod horse when hard is brittle, but a thin shaving from the one wearing tips was as tenacious as whalebone. As I have stated before, I have experimented with all my own horses, X. X. being the one with which the most of the road-trials have been given, though all of them have been exercised more or less on the road. X. X. I have driven long distances—that is, long distances for a road-horse—fifteen to thirty miles in a day, a great part of the time at the rate of eighteen miles an hour. I have driven him fast down quite steep grades, such as are met on the roads in the suburbs of Oakland, and there has never been a bruise on his heel. The wet weather last winter compelled the "working" of the horses on the road, and the favorite ground was Delaware street, from where San Pablo avenue crosses it, to the top of the hill to the west of the University buildings. The distance is probably something over three-quarters of a mile, with a grade of about eighty feet. One of the boys would ride a race-horse I was training, another a trotting filly, a very long strider, in style and action bearing a strong resemblance to Lady Suffolk, while I drove the Alhambra filly to a skeleton wagon or sulky. The race-horse had been injured in his fore legs and his fast work was given up the grade. The filly the boy rode could move through the stretch of the track in forty seconds or better, but on the road she would show a far faster gait. On the track the Alhambra could beat her, but coming down the grade, the other was the faster by at least two hundred yards in the three-quarters of a mile. The boy who rode her weighed 130 pounds, and used a ten-pound saddle. The road was coated with gravel from the beach, though the continual rains kept it somewhat softened, but at times it was hard. She had worn tips from the time I adopted them, and her heels were never bruised. I have never had a horse with bruised heels which wore tips, though I expected trouble of that kind when shoes were taken off and replaced with tips, as the heels were not then in the proper shape. With shoes I have had horses bruise their feet so badly as to suppurate, and, misled by a false diagnosis, ascribed it to gravel, or something else than the true cause. With the feet in the natural condition, it would be almost impossible to bruise so as to injure the sensitive portion in that part of the foot which is supposed to require protection. The wall is not only very thick and strong, but the angle between it and the bars is composed of a material which is nearly as dense, and when the wall is only a trifle lower than the frog, there is an inch of this horny substance to protect the tissues above it. Shoeing, by doing away with the natural functions, deprives this part of the sole of its tenacity, and changes the elastic deposit into crumbling flakes.

CHAPTER VII.

DIFFERENCES OF OPINION IN REGARD TO PROPER ACTION—
EFFECTS OF CHANGES IN SHOEING—PRACTICAL
EXAMPLES—ETC.

The consideration of the action of horses, trotters especially, entails trouble at the outset. It is a subject on which there are such a variety of opinions that it will be almost impossible to take a decided position. One man of great experience favors "high action;" another, equally as well versed, prefers that which the old writers denominated "round;" and still another, whose thorough knowledge of the subject is undoubted, is partial to as little bending of the knee, and the elevation of the hock, as is consistent with sufficient length of stride.

The first will say that the action he fancies is never accompanied with a dilatory motion, and that the energy that compels the superfluous raising of the limbs is a token of speed. The second will claim that there is a happy medium which insures the best results, and the third will call attention to the husbanding of the muscular power and the capacity of the latter to keep up a more uniform rate, retaining the speed when the others lose it. Fortunately, there are two horses in California which represent the extremes, and the two are of such merit that better examples could not be found if the whole country were prospected. Judge Fullerton is a type of the first; Colonel Lewis of the last. Judge Fullerton had such excessive action of the fore legs, that a padded band was necessary to protect the chest from injury from the blows of the fore feet, and though a change in shoeing has obviated the use of this *buffer*, he still brings the shoe in close proximity to the part. His speed is immense. It is doubtful if there is a horse on the track which can show a faster brush, and I have never heard of one which made half a mile in a race as fast as he has shown. I consider that the speed of Colonel Lewis is yet untested. He has shown a quarter in a race at a 2:11 gait, and at this high rate appeared to be "jogging." I will allude, hereafter, to the action of these horses, and, in the case of Fullerton, endeavor to show that the change of shoeing which has resulted so favorably is an approximation of my theory of the benefit of tips, and their superiority to shoes in a case where there is an excess of knee-action. It is more to the point, however, to show that tips have done more to develop the trotting speed in my practice than shoes, and endeavor to prove that such is the logical sequence following the study of the action of the trotter.

I regret exceedingly that my attention had not been directed sooner to this portion of "stable economy." When at Atwood Place, in the vicinity of Chicago, I had so many horses that the test would have been complete ; and from the result of shoes, differing in shape, in weight and in other respects, I am satisfied that had I then known the practical application of tips and toe-weights, I could have done much better. As I noted daily everything in relation to the horses, I can look over my journal and see that I am correct in the statements, without depending on memory, which is proverbially treacherous regarding past occurrences. Every day the exercise was written down, the animals were weighed before going into training, and the effects of every sweat, fast drive, gallop, or run, so far as lessening the weight was concerned, noted. The time of all the fast work and of the trials was registered, and the dates of shoeing and changes of shoes recorded. I had scales in the barn, and employed a blacksmith who had his shop on the place ; in fact, there was everything needed to make whatever experiments I desired, with the opportunity of fully testing the result of them. I had upward of forty horses in training—a few of them with a record, but mostly green horses and colts. The track was a mile, and being a private one, I could keep it in the order I wanted. That I was reasonably successful is shown by the trotting of horses I owned or had in charge. Clara G. trotted and won the three-minute class in Buffalo, in 1871, making 2:30—2:26½—2:26, and could have trotted faster. She won the saddle-race in 2:25, and trotted both without a break. She was prone to leave her feet when I first commenced to drive her, even at a slow gait, though she could trot faster under the saddle, and shoeing had a good deal to do with the change. Ida May had a record of 2:38. She showed ten seconds faster, and would trot through the stretches a 20-gait or better, but I am satisfied that a change in the shoeing to tips would have resulted in a great increase of her speed for the mile. She was a very long strider, and having sprung a quarter-crack—the only horse I had troubled in that way there—she was shod with bar-shoes, though while she wore them she could not trot so fast, and the long stride was lengthened. Nourmahal won a three-minute race in good time, and she beat Lady Mac a dash of five miles in 13:39, when the track was so soft that the foot was buried in it to the depth of the hoof. It had been frozen the night before and thawed out—the worst kind of mud to retard a horse. John H., in a few months' driving, trotted a half-mile in 1:11¼, and sold at auction sale for $7,500, and he has since obtained a record of 2.20 ; and I could extend the list through the "string" with a fair showing of speed in nearly all of them.

The retrospect reveals many troubles which would have been obviated by the use of tips and different-fashioned shoes from those I used at the time, and as it was 1872 before I used toe-weights, until then I was without the benefit of their assistance. It took me months to overcome the tendency in Clara G. to single-foot and hitch in her gait, and during that time I had tried many different shoes. The toe-weight would probably have counteracted the habit in a short time, and I think that a heavy tip would have had the same effect. She trotted very wide behind, and going round the turns she would strike the inside of her hind leg, sometimes as high as the hock, with the edge of the front shoe, and she had to wear shin-boots which extended half way up the hock. After returning from Buffalo she increased her speed very rapidly. Before going there the fastest half-mile she had ever shown on the track at Atwood Place was 1:11½—a quarter in 35 seconds. The next Summer she trotted at several different times the half in 1:08, and once she was timed the quarter in 32¼ seconds, and could always move through the stretches inside of 34 seconds. I could use a lighter shoe than those of the season before, with the edge inside of the horn and rounded, and the shin-boots much lighter, without extending above the hock. She had a habit of lying with her foot under the elbow, and when I got her she had a huge protuberance there, i. e. "capped elbow" which the veterinarian had to dissect out. She wore a roll on her ankle to protect it. The boy who had always taken care of her was so sensitive about it that if any one moved toward her box he would slip in and remove the unsightly appendage, and the consequence was that just before the Buffalo meeting of 1872 he forgot to replace it, and the next morning there was a swelling as large as a croquet-ball. Another operation had to be performed, and she was laid up when I anticipated her showing remarkable time. Had she worn tips this could not have occurred, and I am satisfied that she would have overcome the faulty action much sooner with them, and with still lighter protection in the way of boots.

Albatross I bought the Spring she was six years old. She had a record of 2:48, and two or three months after I bought her I trotted her a race when she was beaten in 2:42. This race took place on Saturday and the next Saturday the same horses were to trot again. She had been wearing an 18-ounce shoe, and a few days before the second race I had it replaced by one weighing 28 ounces and reduced the weight of her hind shoes from 12 ounces to 6. Though a large mare, she was a short strider, and as this was before I had any acquaintance with toe-weights, to remedy it I could only increase the weight of the shoe. She won the second race with ease, trotting in 2:35, and the following week she showed a mile in 2:31, a half mile in 1:13. In 1863 I got Naboclish, with a record of 2:54. He had torn the quarters of his fore feet nearly off, and would cut the leather of the boot as clean as if it had been done with a knife. He had a good deal of knee-action, was very rapid, but was so much afraid of hurting himself that it affected his gait. After driving him awhile I put lighter shoes on in front, took his hind shoes off and rounded the horn at the toe so that when he struck his quarter the blow would not be felt, the boot offering ample protection. It gave him confidence at once. At that time I had a training track on my farm in Iowa, the distance around it being five-ninths of a mile. I got him in April; in August he trotted three times around the track, one mile and two-thirds, at a 2:23 rate, and I won fourteen races with him that Fall without losing one. During this campaign he never had a shoe on his hind feet, and though he trotted on many hard tracks, the horn did not break, nor did he show any tenderness. Of course it is entirely conjectural what the effect of tips would have been on these horses, or whether toe-weights would have accomplished more than the heavier weight in the shoes. But, reasoning from analogous circumstances, there can be little doubt that great advantages would have followed their use. In the case of Naboclish it proved that the bare hind feet could withstand the hardness of the tracks, and in place of the 28 ounces on Albatross, it is almost certain that a tip and toe-weight of 14 ounces would have been more effectual. I am satisfied that with their aid the defects in the gait of Clara G. could have been remedied in one quarter of the time it took, and that the injuries to the elbow and hock would have been avoided.

The continuation of the history of the work given the Alhambra filly will show what was the result in her case. Her action was such that I did not expect that the tips would do as well as those horses which had more knee-action and a longer stride, and so I gave her a more thorough trial with shoes on than if her gait had been different. As I have stated before, I ascribed the shortening of the stride to driving her when she was slightly lame, and to her running a great deal when turned out in the field before she had recovered. She wore shoes from August 16, 1877, to November 5th, then tips to December 10th. April 13th, 1878, I put copper tips on, reaching across the toe, and on May 21st replaced them with light shoes with very low heels, so as to give a better fastening for a quarter-boot. She wore these shoes until June 3d, when they were replaced with 16-ounce ones with the "Eclipse" fastener for toe-weights. She wore that weight until July 27th, when I put on tips weighing 6 ounces each with the same weight-fastener, and these she wore until August 18th, when she sprained her hind ankle slightly and was again turned out. The fastest mile she trotted when wearing the shoes from August 16th, 1877, to November 5th, was 2:52. The result of the 8-ounce tips was that on November 8th she trotted a quarter in forty-one seconds, the mile in 2:48¼. Still I was induced to try the shoes again on December 10th, thinking that her heels were hardly in proper shape for the tips, and the horn being of slow growth from wearing them so long, the shoes might be better. Another, and more forcible reason, was that I had made a present of her to my friend B. C. Wright, and I did not want my preference for tips to mislead me with the property of another, and whenever I was in doubt regarding the effect, I would stick in that case to the established custom. Mr. Wright was perfectly willing that I should try the experiments, and from what he had seen of the benefits of the tips to the feet, he is using them on the horses on his ranch in Mendocino County, and finds such an improvement that he says he would not have them replaced with shoes if the blacksmith would do his work for nothing.

The favorable result of the 28-ounce shoes on Albatross determined me to try the same plan with Avola, and each front shoe weighed 22 ounces. While wearing them she was driven on the road, the wet weather making the track unfit to work upon, but as the other filly could beat her so easily, I think that there was no improvement while she wore them. The copper tips extending across the foot, the whole portion anterior to the frog being covered, appeared to enable her to do better, but as she kept cutting and bruising the quarter of her near fore foot, and, owing to the want of the heel of the shoe, the quarter-boot could not be kept in place. The shoes put on the 20th of May were very thin at the heel, so as to allow the frog to touch, and with them I used "Eureka' toe-weights—shoe and weight, 14 ounces. The 22d she moved through the stretch in forty seconds, striding farther than before, which I gave the toe-weights the credit for accomplishing. But with these shoes she was inclined to hitch at times, and I replaced them with others weighing 16 ounces on June 3d, and the inventer, H. L. McKinney, having sent me his "Eclipse" weight and fastener, I attached the fastener to the shoe. There appeared to be an improvement, as on the 8th she trotted the quarter in thirty-nine seconds—the toe-weights 9 ounces, making 25 ounces on each foot. On the 10th of July she made the same time, and, until she became accustomed to this weight, I thought it best to restrict her fast work to short brushes, for fear of injury to the legs. On the 24th she trotted half miles in 1:23—1:21½, and on the 27th three-quarters in 2:05—2:03¼. The 1st of July I drove her three heats of a mile, the time being 2:50½—2:49—2:49¼. The work was continued through the month, and on the 24th of July she trotted a mile in 2:47, the fastest she ever made with shoes on.

On the 27th of July I put tips on her front feet weighing 6 ounces each, attaching the "Eclipse" fastener. This gave me the power to keep quarter-boots in place without the heels of the shoe, the iron in front holding the strap, which came round the heel so that it could not rise. The pattern of quarter-boot was the one which is something like a "tunnel" boot, with a strap going round the foot to keep it from getting out of position. The 30th of July I drove her a mile in 2:44¾. She had the same toe-weight that she wore with the shoe, and there was a gain in a few days of 2¾ seconds. On the 4th of August I drove her a mile in 2:43¾, and on the 8th in 2:44½. This established in my mind the fact that she could trot faster in a 6-ounce tip than in a shoe weighing 10 ounces more; though from wearing the shoes so long, her heels were not in the best shape for the tips. There was another thing which caused me a good deal of trouble. Her upper jaw is a trifle wider than the lower, and this entailed a sharp edge on the "grinders" above, and this edge had wounded the inside of the cheek so that she carried her head much to the side, with a tendency to pull hard on one rein. I filed the teeth as well as I could, but the injury was so far back that it was difficult to get at the seat of the trouble, and it would require an expert, like Dr. House or Dr. Clark, to make it effectual. Still, the rude dentistry helped her, and she drove straighter after the operation. On the 18th I checked her higher than usual, and the result of that was a slight sprain of the hind ankle, not sufficient to cause lameness, though there was heat and swelling of the joint. I turned her out, after fomenting the part for a few days, not only on account of the injured ankle, but the run would give her mouth the opportunity to get entirely well, the heels would grow down, and as I would be absent several weeks at the fairs, I did not want her to have any work I did not witness.

The other filly, Columbine, alluded to in former papers, I sent the 9th of March to breed to Governor Stanford's Electioneer, so that Avola was the only trotter experimented with during the summer. It may appear that with no greater speed than a mile in $2:43\frac{3}{4}$ this was not much of a test, but there are plenty of young trotters which never show as well, and if there is a decided improvement in this rate, it is a fair inference that there would be a corresponding increase in that which is faster. In all of my experiments with tips on trotters, I have been restricted to three animals, of the same age, and of something the same breeding. All of them have shown faster than a 2:40 gait for a quarter of a mile, when wearing tips; and one—San Diego—when a four-year-old, made that distance in thirty-seven seconds. With a number to choose from, to obtain three untried colts which would make so good a showing would be held very fair, and without opportunities for selection I consider the performances are far above the average.

Columbine had very little work in comparison with Avola, and the latter has only had a small portion when compared with other colts. The greater part of my time being occupied with my regular business, I could not give the attention to the education of the horses which is such an absolute requirement for them to do well, and hence there is another drawback to take into consideration. At Atwood Place the trials of one Summer would have given me more information than ten years of limited experiments, and to obtain the results obtained from such a meagre list, strengthens the belief that with fuller exemplifications the benefits will be still more apparent.

Before closing this chapter it may be as well to refer again briefly to Judge Fullerton. His action is considered as nearly perfect as can be when there is a sharp bending of the knee, and the veteran trotter has certainly proved that he has strong claims to be considered one of the foremost of the celebrities of the tracks. Until he came to California he carried a fore shoe weighing nineteen ounces, and then he had to wear a breast-pad to keep him from pounding his chest. Under Hickok's charge the weight has been gradually reduced until that of the front shoe is twelve and a half ounces and the hind one six ounces. It was conceded that he faltered somewhat in finishing the mile with heavier iron, while at San Jose and Stockton it seemed as if he had a world of speed left, and came from the distance home with as much energy as he exhibited when he took the lead around the first turn. The pad on the brisket is not required, and altogether the reduction has been beneficial. He picks his fore feet up so truly, and places them so squarely on the ground, that a pair of front shoes will wear him a long time, while nearly every two weeks a new set are made for the hind feet. The bar-shoe is used in front, and this partial bearing on the frog may have a good influence. Notwithstanding his seven years of trotting in many races per annum, I am of the opinion that he could have trotted three faster heats at San Jose than ever before in his life, and to the difference in shoeing may be ascribed much of this improvement.

A portion of this was due to the lightening of the iron, a part to the "frog-pressure" arising from the bar-shoes. But I am well satisfied that tips would have been still more beneficial, notwithstanding Fullerton's feet have thin walls, and all of them white in color. It may be prejudice which ascribes a greater weakness in white horn than is found in the darker shades, though the idea is so general that it has probably good grounds to sustain it. White horn is more easily cut with the knife, and easier broken when a shoe is worn, but as Naboclish's hind feet were white, and they endured fourteen hard races, and training from Spring until Fall, there is little question of the heels standing when the toe is protected from wear. In all probability a very light tip would have been found sufficient in a horse with so much knee-action as Fullerton, and one weighing from four to six ounces enough for him to wear. Doubtless the heavy front shoes, of nineteen ounces each, were worn to extend his stride, but as that was not shortened by the reduction to twelve and a half ounces, a further subtraction would have had no ill effect. The race-horse strides further with plates on his feet than when he is wearing training shoes, and though the weight gave Albatross a longer stride, it was from a directly opposite reason, viz: a lack of knee-action. The natural super-exertion, in Fullerton, was heightened by the heavier shoe, and, as his stride was not curtailed by a contrary course, the inference is just which contemplates a still more radical change. And then comes the question of the effect of the weight on the heel, which the bar-shoe gives, in contradistinction to the weight on the under part of the toe, in the tip, or on the front part of the wall, in the toe-weight; but the consideration of these abstruse problems will come in more appropriately hereafter.

CHAPTER VIII.

ENDORSEMENT OF TIPS—THREE-QUARTER SHOE.

The following letter from Dr. Taliaferro, I take the liberty of publishing, though that was not the intention of the author; and, while many have been received of like tenor, so far as an endorsement of the use of tips, there are objections in this which require an answer:

SAN RAFAEL, November 26th, 1880.

MR. SIMPSON—*Dear Sir:*—I am glad we are going to hear more from you in regard to shoeing, and particularly in relation to tips. Everything you write on that subject, and indeed on every subject pertaining to horse matters, is very interesting. I used last winter on my riding mare the tips put on in the manner recommended by you about a year ago, *i. e.*, cutting away the crust until the shoe is upon a level with the sole of the foot. It answered admirably during the winter and spring months, but when the dry season sets in our roads become very hard, and my mare wears away the frog and heel as if they had been ground on a grindstone. Besides riding, I drive the mare a great deal, and, whether riding or driving, I use her very fast. She is thoroughbred, and the hardest animal on the foot and shoe I ever saw. I have been partial to the tip for a good many years, and went at them with renewed vigor after reading your article some two or three years ago on the subject. On a Monday filly I have; I encountered the same difficulty in the dry season, and have had to give up the tips by reason of lameness caused by tenderness of the heel. I shall always use the tips, however, put on as recommended by you, every winter, for, if for no other reason, it leaves the foot in such splendid condition to go through the summer with. I have not seen that any one else has made any report to you concerning this matter, and have simply mentioned my experience in order that you may take it for what it is worth.

ALFRED W. TALIAFERRO.

The subject is so important that, in my estimation, it overshadows any other part of stable economy, or any one phase in the management of horses, especially young colts. I am fully satisfied that the only drawback to the training of yearling trotters is the fancied necessity of shoes. When thoroughbreds are galloped at that age, they are left unshod, and hence there is no injury, but the impression being so general that the weight on the fore feet is an imperative necessity with the young trotter, shoes are worn. Until that is removed there will be serious injuries following the use of the full shoe. At this period of life the foot grows more rapidly than at any other age. Hampering that growth is sure to be followed by a structural change, which can scarcely be remedied in after life. Only a few days ago I saw a number of yearlings which had been shod for months, and two-year-olds which had worn shoes for a year or more. A glance disclosed that the feet were not in normal condition, every one having more or less of a cramped look. Few are aware of the "spring" there is to the foot of a horse which has never been shod, compared with the rigid box of the animal that has worn shoes. The instrument, which I used for a time to prepare the foot for the tips, discloses this. It is clamped on the foot, a small lever giving the pressure, and though the power is only sufficient to hold it in place, the elasticity is obvious. Applied to the shod foot there is none of this yielding, and a notch or two in the ratchet, which holds the lever, is all that it can be moved, whereas it passes over quite a space in the other. The reference so often to X X is compulsory, inasmuch as the experiments in the last two years have been mainly confined to him. His foot measures a little over five inches transversely, the length five inches. This is the natural—or nearly so—proportions of the foot, though in a horse which has never worn shoes, the transverse diameter will still more exceed the longitudinal, the base being very nearly a true circle. Having a set of shoes worn by Santa Claus, I made a comparison of these and the foot of X X. This must be prefixed, however, with the statement that Santa Claus has an extraordinarily good foot, and, for a horse which has worn shoes, it is in remarkably good condition, no doubt owing to the extra care taken in his shoeing. The transverse measurement of the shoe is 4 13-16 inches; it is 5 7-16 long. Doubtless the horn projected a trifle over the sides, and did not come flush with the heel. From the marking on the iron it came within 3-16 of the heel, which would leave $5\frac{1}{4}$ inches, though in all probability it was not set flush with the toe. The foot of X X, resting on a paper, made an impression which a circle entirely enclosed. The imprint of the shoe of Santa Claus showed an inch posterior of the circle struck from a center the same distance in the rear of the toe as one-half the transverse diameter. This proves the structural change incidental on shoeing, and when a naturally superior foot, supplemented with the very best care, exhibits this departure from the true form, what must be the effect on the foot of a yearling, hampered with the full shoe at the most active period of growth? Something akin to that which the infant shoe, continued on the ladies of high birth in China, during the period of growth, displays, with the deformity as much pronounced in degree. The tip, properly adjusted, does not interfere with the growth in the least. Placed anterior to the widest diameter, the growth is not hindered, and the horn is at full liberty to expand in every direction. On Wednesday, the first of this month (December), the off fore foot of Anteeo was placed on a paper, and the outline accurately marked. This was bisected with a line from the center of the toe to the heel, and midway of these points a right angle was struck. The length was $4\frac{5}{8}$ inches, and the width was exactly the same. A circle from the point at the intersection of these two lines struck the toe, outside quarter, and the heels, but came a little outside of the inside quarter; otherwise the foot was nearly round. The disparity in the growth of the different sides of the foot is partially owing to natural causes, partly to individual characteristics. As all men who have given attention to the horse's foot are aware, the horn on the inside quarter is thinner than on the outside of the foot, and the formation of the wall straighter on the inner.

Anteeo throws the most of the stress on the outside, as is proven by the greater wear of the tip on that side, and in this respect he is like a large majority of the horses we have had. That a colt nineteen months old should have a foot within half an inch in width of one which is nearly nine years old, who is a hand taller and three hundred pounds heavier, is positive proof of how soon the hoof gets its growth, and the danger of restricting that growth at so critical a period. Although Anteeo has only worn hind shoes from August 6th to November 10th, with an interval in September when shoes and tips were pulled off, it is perceptible, on close scrutiny, that even that short time had a prejudicial effect. On November 10th, the shoes were replaced with those which merely covered the toe and the outside quarter, and from hence these are the only kind he will wear as long as he is under my control, or until I see good cause for forsaking them. The wear on the outside of the hind feet of both he and XX is still more pronounced, and either will wear a full shoe completely through on that portion when the inner is intact. These three-quarter shoes are the only kind which XX has worn for some time, and they give perfect satisfaction so far.

The objection which the correspondent has to the wearing of tips in the summer time can be overcome with the use of this kind of shoe, and in this chapter a cut is given showing the kind tried on the fore foot, with marked success. Still, I am satisfied that no matter how hard the roads are, with due preparation of the foot, tips can be worn at all seasons. In Oakland the winter is the most trying time. The Oakland streets are constructed of broken stones, with gravel or crushed rock for the superstructure. The country roads are mainly coated with gravel, and, when wet, the attrition wears the horn more rapidly. It is like the application of water to the grindstone, and that when the horn is softer from the absorption of moisture. When tips were first used, I did not anticipate that it would do to drive a horse on the Oakland streets and the drives I most fancied constantly, and though convinced, from the first, that for track purposes, and on ordinary country roads, they would meet with all the requirements, I would not have been disappointed had they been found wanting after so severe a test. As stated heretofore, XX has only worn two pairs of front shoes in five years. He has been in constant use on these streets and roads, and latterly he has been given long and rapid drives with scarcely a day's intermission. The wall and bars are worn nearly level with the frog, though there is plenty of horn, and the frog is full. There is not a particle of discoloration, or indication of bruising; his feet are cool, and the tendon which was injured when wearing one set of shoes, at the time "Spanish Charley" had him in training at Sacramento, is now as "clean" as it ever was, and the most skillful veterinarian would fail to decide which leg had met with the injury.

The three-quarter shoe will entirely obviate the difficulty Dr. Taliaferro mentions.

The illustration is a copy of one of a pair of shoes made after a pattern sent to William Zartman, Petaluma. The intention was to use it with the toe-weight which Mr. Zartman has patented, and by getting them made at headquarters an assurance was given that the spur would be properly made, and the shoes such as were wanted. As the various kinds of toe-weights, their uses and abuses, will form a distinct chapter of the treatise, it is unnecessary to say more at present than that the Petaluma is meritorious, equaling the best in nearly every respect. The objects in using shoes of that pattern on the forefeet were, in the first place, to remedy a faulty method of picking up the foot, to see if it allowed the proper elasticity at the heel, and should it be proven that tips could not be used on streets and roads like those of Oakland, it might take the place of them, obviating the greatest drawback of the full shoe. The experiments were satisfactory on all these points. The peculiarity in action consisted of the horse lifting the forefoot with a kind of a twist, in which the inner side was elevated more than the outer, and, on the hypothesis that unequal weight acts in carrying the foot towards the loaded part, it was expected that the defect would be remedied.

Nearly twenty years ago I experimented with a pacing horse, and tested this theory until satisfied of the truth of it. This was described in "Horse Portraiture," though at that time I had never heard of the application of weight to the horse's foot in any other method than in the shoe, or the loaded quarter-boot. The pacer was a most inveterate "knee-knocker," and by making the outside of the shoe very much heavier than the inside, and using a contrivance placed in the "fork," he was cured of the propensity. It is certain, however, that a greater effect is obtained from the weight being placed higher up than the sole of the foot, though in many instances the lower application of it will be found sufficient. With the first use of toe-weights it was thought that the advantages came from the weight entirely, and hence horses' legs were ruined by carrying loads which permanently injured the tendons. Now it is demonstrated that it is the inequality of distribution which gives the best effect, and that a 6-oz. weight on a shoe of 10 oz. is as effectual as 16 oz. on one which weighs $1\frac{1}{4}$ lbs. There being an entire absence of weight on the inside quarter, that on the outside had a more potent effect. "Side-weights" on the fore feet are inadmissable, as there would be greater danger of injury to the hind legs when put on the outside, or the knee and arm if the excrescences were on the inner. The toe-weight can be given an angle in either direction, though there are serious objections to using it in any other than a straight line. Thus I was compelled to substitute something akin to the former experiences, and, as will be noticed in the cut, the heel of the shoe was made wide enough to cover the whole space between the frog and the wall. It was made thin in order that the foot might be kept level, as well as permitting the frog to perform its functions. Only one side was fettered, and even that free in comparison with the rigid embrace of the full shoe. It is well known that, owing to the thinness of the horn on the inner side of the foot, there is a far greater degree of elasticity in that portion, and even if the outer was somewhat bound, much of the injurious result of contraction is obviated. Much stress has been laid on the propriety of restricting the nailing to the anterior part of the wall, and it is certainly a better plan than to drive them further back. The friction, however, between the wall and the iron soon wears a depression, and this prevents the expansion in a great measure. With the full shoe the two grooves are nearly equivalent to back nailing, and the posterior part of the wall is held together as though it were locked in a vise. A glance at the engraving will show that the weight of the animal, when thrown on the foot, will have the effect of increasing the expansion of the inner quarter, and the delicate mechanism enclosed in the horny covering will have nearly as much room as in the natural state. The frog is employed, and the labor keeps it in health. This has been the savior of the Goodenough shoe, and has kept it in use notwithstanding the serious defects which mar it. People sneer at the preface to the advertisement, and chuckle over the alliterative phrase, "No frog, no foot; no foot, no horse." But this has been the redeeming trait, and the inventor is worthy of great credit for insisting on permitting this important organ to do the duty it was intended to perform. Mr. Goodenough came very near adopting the right method, and had he battled with the same persistency for a still closer adherence to nature he would have accomplished more than he has. Perhaps not. In that case he would have met with a greater degree of stubbornness, and the whole rejected, whereas it is about the only improvement which has had any favor; it has held its own in the face of ignorant opposition, and opposition which was characterized by scenes which were strangely in contrast with any degree of common sense. In Chicago, some eight years ago, many of the journeymen blacksmiths refused to nail them on, and they paraded the streets with banners and mottoes to intimidate their "bosses" from using them.

At that time I had several long conversations with Mr. G., and gave the shoes a trial, being in the main pleased, though the cutting of the channel so deeply I was opposed to, and endeavored to remedy that by having the shoes flattened at the heels. Among other things which were discussed, was the cause of corns, and he very correctly ascribed the origin of them to the pressure of the ordinary shoe. David A. Gage had a fine road-horse which was grievously troubled with them, and at Mr. Goodenough's suggestion tips were put on him. He went much better, but owing to the erroneous method of preparing the foot they had to be given up. Had Mr. Goodenough had the channel, which he directed to be made for the reception of his shoe, cut only far enough back to receive the tip, allowing the whole of the posterior part of the foot to rest on the ground as well as the frog, there would have been but little variation from the treatment recommended in these papers. It is rather surprising that he did not come to the conclusion, and that if there were so many benefits to be derived from "frog-pressure," there must have been analogous advantages following the other natural provisions against the injuries arising from concussion and contraction. The trouble was that he overrated the beneficial effects of frog-pressure, and in estimating that it would make amends for all the evils which follow shoeing, he did not progress to the ultimate point. His shoe is a thick one, requiring so deep a cutting away of the horn that the continuity between the wall and sole is weakened, and had it not been for the jar being so much weakened by the frog, the animal would have been lamed in a brief period.

Many think that the wall and sole of the horse are the same, and that the difference in texture is owing to something they never troubled themselves to discern. Though intimately connected, they are entirely different, and maceration will separate them in a short time. It is obvious that when the junction is made so much thinner, especially at the weakest point, injury must arise from the cutting away of the horn of the wall and sole to such a depth as is necessary to imbed a thick shoe so that the ground surface is on a level with the frog. In the instances which Dr. Taliaferro gives, the probable cause of the soreness was the wearing away of the outer side of the foot until it is something like the channeling which the Goodenough process directs. Finding that such a large majority of horses wear the outside the most, without direct examination, the assumption is probably correct. The gliding motion, which is a peculiarity in the action of the blood-horse, entails greater friction and greater wear; and the feet having been pared during the time when shoes were worn, the growth of one winter is not enough to withstand the work of the summer. The three-quarter shoe will remedy this, and the benefits which arise from the whole foot being unfettered for one-third of the time will enable the animal to wear this form of shoe with good results. It is so much better than the ordinary kind that it does not require long arguments to prove the superiority. The setting must be the same as the tips, the shoulder square where it ends on the inner side of the toe, and the inner side of the foot left flush with the ground surface of the iron.

There is a prevailing opinion that low heels increase the strain on the tendons of the fore legs, and this is so generally entertained that it has been reiterated over and over again in arguments against the use of tips, and generally in a confident, dogmatic way that was to end the argument at once. Veterinarians are nearly unanimous in recommending that the heels be raised when an animal is suffering from a strain of the back tendon. The propriety of that treatment we shall not question, at least in this article, but attempt to show that there is less danger of "breaking down," or severe strains when the foot is low enough at the heel to allow the parts to perform their natural functions. The muscles are the main motive force in animals, and this arises from the power they possess of dilatation and contraction. In some parts of the frame they act directly, in others, through tendons. These are firm, compact bands of white bundles of fibres, very nearly insensible, though the sheathing or covering is acutely sensitive when there is inflammation. In a former article (page 33) there was a cut showing the cannon, sesamoid, pastern and coronet bones. When published before, the object was to show that a shoe which projected behind the heel was injurious, the cut and the comments being copied from Miles' essay. He very forcibly explains why the long shoe was bad, and if so bad, the argument was still stronger for leaving the posterior part of the foot as nature made it. The high heel has the same bad effect as a long shoe, with the addition of being still worse from placing the bones in a more upright position, the baleful influence extending to when the animal is not in motion. The cut does not show the whole of the coronet bone, as Miles terms it, though it is usually called the lower pastern bone; below that is the navicular or shuttle bone, and the coffin or pedal bone. The front part of the latter retains about the same obliquity as the two above it. The navicular is a transverse bone, and the union of three form the joint which is the most susceptible of injury of any in the frame of the horse, not even excepting that of the hock. The tendon passes under the navicular bone, resting upon the sensitive frog, an inch or more back from the point. It is evident that if the heel is raised more than is natural, that the angle will be more upright, and a wrong "set" given to the navicular joint, and this is followed by the upper pastern bone, and again by the cannon, in order to restore the harmony of position which has been disturbed. This can only be carried as high as the knee, which is "sprung" forward to relieve that joint as much as possible. Therefore, it is palpable that the elevation of the heel gives a wrong placing of the bones, and this must exert an influence on the tendons and ligaments, which is injurious. All are aware of the exhaustion which results from keeping one position, especially if that be an unusual one.

The arm held at right angles from the body can only be sustained for a short time, and to stand in a rigid posture can only be endured for a few minutes. It may be that the trouble begins long before the actual injury, and the constraint arising from the abnormal placing of the feet and limbs is the commencement of the difficulty.

As has been stated in prior articles, experience has shown that in every horse of mine which has given way in the tendons, there has been something wrong with the feet. Very trivial, perhaps, and only noticeable after the strictest scrutiny, yet not in a truly natural condition. A slight contraction, narrowing and hardness of that part of the frog which touches the ground, a little extra heat perceptible, apparently a trifling ailment which would be entirely overlooked by a person who was not familiar with an actually natural foot. Still, if one foot was affected, that leg was the first to yield, and though a horse might break down with the best of feet, the conjunction is worthy of notice.

It is not necessary to enter into a minute description of the tendons of the fore leg, being sufficient to state that they are acted upon by the huge muscles which clothe the shoulder, and follow the radius until connected with the tendons near the carpus. There is a tremendous force in these massive bundles of fibres, and this power is needed to send the animal along at such a pace as the fast galloper can make. The old idea, not yet surrendered by ninety-nine in a hundred of horsemen, that all the fore legs had to do was to support the body, while the hind propelled it through the air, is effectually exploded by the instantaneous pictures. The last supreme effort before the horse is hurled through the air is made by one fore leg, and consequently there must be a corresponding energetic movement to effect the purpose. The foot has been placed on the ground, nearly on a line with the nose, and the other foreleg does its part by carrying the weight along until it is under the brisket. When so far back that a vertical line from the toe will strike behind the cantle of the saddle, the grand propulsive force is applied. As the body is carried over the foot, the pastern is bent until the ankle touches the ground, and at this point the injury from the high heel or the long shoe comes. It is evident that the tendon which passes under the navicular bone is forced to a sharper angle, and when the contraction of the muscles act on the cords, a greater resistance is met. By looking at the cut, and imagining the slight angle which the bones form with the ground surface changed to one which is below a parallel line, some idea may be formed of the increased resistance to be overcome. When the tendon passes over a joint, it has a groove to run in, this being lubricated by a mucilaginous liquid called the synovial fluid, in common parlance joint-oil.

Should the cord be pressed with a greater degree of force than is necessary, it is like applying a brake to a wheel, the high heel being as effective in performing this as the Westinghouse on a railroad train.

Short bulky muscles can exert a more potent contractile force than those which are longer, but the thinner and elongated fibres can accomplish more extensive movements. This is the difference between the part bred quarter-horse and the thoroughbred. The former goes into his stride at once, the quick action of the bunched up muscles in arm and gaskin quickening the action, while the longer leap is deliberate, and the animal is scarcely well settled in his stride when the other is exhausted.

There has been a great deal of controversy over the question of the greater frequency of break-downs in the long distance runner in comparison with the short horse. Both are shod similarly, and the immunity of the latter would appear to be a strong argument against the claim that the elevation of the heel is detrimental. This does not follow.

The quarter-horse has little work in proportion, and his races are not so violent a strain. He runs on a hard track, and this is additional proof that it is the effort to rise which injures the tendons. The deeper the sinking of the feet the more violent will be the muscular contractions, and the greater strain on the cords which convey the power.

Veterinary writers agree that one of the surest symptoms of navicular diseases is the "pointing" of the afflicted foot. If one only is diseased, it will be constantly thrust forward ; if both, there will be a change every few minutes from the intolerable pain when the foot is under the body. This does not arise so much from the weight as from the position. The further forward the foot is placed, the more resemblance there is to the low heel, and the similar posture affords relief. Cases of navicular disease are extremely rare, excepting where the heels are high, and those we have seen are almost invariably marked by severe contraction. A natural foot, one entirely free from disease, and which has never been shod, presents at the heel a very different appearance. The base line is formed by two curvatures of the inner and outer quarters, and these are joined by a reverse curve representing the posterior part of the frog. If the animal be permitted to run at large, or is worked without shoes, there will be no change. In a previous paper something of the same ground was covered, though the long periods intervening, awaiting the result of experiments, has compelled reiterations. This desultory method of treating the subject has not been without advantages. Every point has been covered by practice, so far as it has been practicable to make the test. The few horses which we have had in the last two years has extended the application, and left things which might have been proven, partly conjectural. To establish the theory that there is far less risk of breaking down from the feet being unhampered, would require more actual results than we now possess, and though firm in the belief that such will be ultimately proven, there is not sufficient data to state it authoritatively.

As many of the Eastern trainers of race horses have discarded shoes, a comparison could be instituted there, and here there is one instance. The horses in the Santa Anita stable have been trained barefooted, and with the exception of Clara D., who had shoes put on to make a long railroad trip, the only incumbrance has been the racing plates. There were less casualties among them than in any stable on the coast, and they certainly did their share of work. Mr. Martin informs me that he is so well convinced that it is preferable that he will continue the practice. Clara D. was shod without his knowledge, and it is not likely such a course will be pursued again after the way he emphasized his disapprobation.

The Australians have forsaken shoes, and the late William Dowling informed me that they had found it to be far preferable to the old method. When in races they are plated in order to give them a "hold" from which to spring, and as the courses there are coated with grass, there may be a necessity for a catch. The tip, however, would afford this just as effectually as the full plate, without the danger of twisting from the heel.

CHAPTER IX.

GUARDS AGAINST CONCUSSION.

In all animals which have to progress with any degree of celerity on pedal extremities, nature has provided guards against the injurious effects of concussion. In considering the proper manner of keeping effective the natural provision in the foot of the horse, a brief comparison with other quadrupeds will not be out of place. In all are contrivances, admirable and complete, in a state of nature, for the intended uses, and valuable lessons will reward the investigator in this department of animal economy. In former papers I have alluded to the foot of the greyhound, and now the antitheton of that fleet courser will come under review. There could scarcely be two animals selected which are more dissimilar than the greyhound and elephant—the former a type of grace, agility, lightness, the other massive in its hugeness—so ponderous as to awaken a feeling of awe in a' person who has not been accustomed to seeing them.

The skeleton of the greyhound differs from that of the horse, in there being a sharper angle between the shoulder-blade (scapula) and the bone of the upper arm (humerus). From the elbow to the knee is far longer in proportion in the dog, and the cannon, or metacarpals, correspondingly shorter. In place of the lower pastern shuttle or navicular bone, and the coffin bone, the dog has digits similar to those of the human hand, and these are in nearly a horizontal position. The spring of these digits, and the cushion of the sole, together with the angle of the scapula and humerus, are effective guards against the jar which follows the striking the earth after the bound. I do not intend to go into an analysis of the stride and action of the greyhound until I have studied the last series of photographs taken at Palo Alto. That it will vary from that of the horse, I feel confident from the configuration being so different; but without that light there would be just as much liability to error as there has been in regard to the action of the horse in a fast gallop. More so, as there is fully as much speed, with quicker movements, so that the brain is still more troubled to retain the picture which is yet more transitory. For the following account of the elephant we are indebted to the *Asian*, a newspaper published in Calcutta, India, and as a chronicle of "sport, shikar, the turf, garden, tea, indigo, whist, chess, etc.," full of interesting matter. Published in a country where the elephant is utilized, dependence can be placed on its accuracy, particularly as the author, R. A. Sterndale, F. R. G. S., has been a resident of India long enough to give him an intimate knowledge of the subject. As it is only a small portion of the essay which we have pirated, and bearing as it does so intimately on the topic under consideration; and then, again, being so far away, the acknowledged capture may be pardoned:

The elephant has seven cervical vertebræ, the atlas much resembling the human form; of the thoracic and lumbar vertebræ the number is twenty-three, of which nineteen or twenty bear ribs; the caudal vertebræ are thirty one, of a simple character, without chevron bones.

The pelvis is peculiar in some points, such as the form of the ileum, and the arrangement of its surfaces resembling the human pelvis.

The limbs in the skeleton of the elephant are disposed in a manner differing from most other mammalia. The humerus is remarkable for the great development of the supinator ridge. "The ulna and radius are quite distinct and permanently crossed; the upper end of the latter is small, while the ulna not only contributes the principal part of the articular surface for the humerus, but has its lower end actually larger than that of the radius—a condition almost unique among mammals."—*(Prof. Flower.)*

On looking at the skeleton of the elephant, one of the first things that strike the student of comparative anatomy is the perpendicular column of the limbs; in all other animals the bones composing these supports are set at certain angles, by which a direct shock, in the action of galloping and leaping, is avoided. Take the skeleton of a horse, and you will observe that the scapula and humerus are set almost at right-angles with each other. It is so in most other animals; but, in the elephant, which requires greater solidity and columnar strength, it not being given to bounding about, and having enormous bulk to be supported, the scapula, humerus, ulna and radius are almost in a perpendicular line. Owing to this rigid formation the elephant cannot spring. No greater hoax was ever perpetrated on the public than that in one of our illustrated papers, which gave a picture of an elephant hurdle-race. Mr. Sanderson, in his most interesting book, says: "He is physically incapable of making the smallest spring, either in vertical height or horizontal distance. Thus a trench seven feet wide is impassable to an elephant, though the step of a large one, in full stride, is about six and a half feet."

The hind limbs are also peculiarly formed, and bear some resemblance to the arrangement of the human bones, and in these the same perpendicular disposition is to be observed; the pelvis is set nearly vertically to the vertebral column, and the femur and tibia are in almost a direct line. The fibula, or small bone of the leg, which is subject to great variation amongst animals (it being merely rudimentary in the horse, for instance), is distinct in the elephant and is considerably enlarged at the lower end. The tarsal bones are short and the digits have the usual number of phalanges, the ungual, or nail-bearing one, being small and rounded.

Another thing that strikes every one is the noiseless tread of this huge beast. To describe the mechanism of the foot of the elephant concisely and simply, I am going to give a few extracts from the observations of Professor W. Boyd Dawkins and Messrs. Oakley, Miall and Greenwood: "It stands on the ends of its five toes, each of which is terminated by comparatively small hoofs, and the heel bone is a little distance from the ground. Beneath comes the wonderful cushion composed of membranes, fat, nerves and blood-vessels, besides muscles, which constitutes the sole of the foot."—(*W. B. D.*, and *H. O.*)—of the foot as a whole, and this remark applies to both fore and hind extremities; the separate mobility of the parts is greater than would be suspected from an external inspection, and much greater than in most ungulates. The palmar and plantar soles, though thick and tough, are not rigid boxes like hoofs, but may be made to bend, even by human fingers. The large development of muscles acting upon the carpus and tarsus, and the separate existence of flexors and extensors of individual digits, is further proof that the elephant's foot is far from being a solid, unalterable mass. There are, as has been pointed out, tendinous or ligamentous attachments, which restrain the independent action of some of these muscles, but anatomical examinations would lead us to suppose that the living animal could at all events accurately direct any part of the circumference of the foot by itself on the ground. The metacarpal and metatarsal bones form a considerable angle with the surface of the sole, while the digits, when supporting the weight of the body, are nearly horizontal.—(*M. and G.*)—This formation would naturally give elasticity to the foot, and, with the soft cushion spoken of by Professor Dawkins, would account for the noiselessness of the elephant's tread.

On one occasion, a friend and myself marched our elephants up to a sleeping tiger without disturbing the latter's slumbers.

It is a curious fact that twice around an elephant's foot is his height; it may be an inch one way or the other, but still sufficiently near to make an estimate.

It is evident that there must be provided something more than is found in the foot of the dog and the horse to counteract the force of the immense weight sustained by the massive perpendicular columns. The cushion, composed of the tissues the best adapted to give softness and elasticity, is spread out to an immense size to increase its effectiveness. "Twice round" the foot of a horse would lack several inches of reaching the elbow, and in a majority of thoroughbreds the proportion would be nearly as one to five in measuring the height. It is also evident that the smaller foot is one of the elements of speed, and to prove this it is only necessary to compare that of the draft horse with the pedals of the racer. The foot of the dog is much larger in comparison with the size, especially the weight, and hence a softer cushion is admissable. To withstand the increased attrition, the material must be harder, and in lieu of the indurated sole of the foot of the dog there is the still firmer frog. But, as has been partially explained heretofore, the frog is a reinforcement, aiding the parts which have stood the hardest shock of the battle. The first charge has been met with the wall, and the forcing the wall apart, made possible by the commissures, gives the frog the opportunity to play its part. It is astonishing that this should be so persistently ignored, when all but one crazy theorist, and perhaps a few of his disciples, admit the importance of allowing the frog to perform the part Nature intended.

A general who would shut up his reserves in a space where there could not be a chance for the least exercise, who left them there until the muscles were wasted away and the nervous force entirely lost, who went into an engagement in which he was defeated entirely from neglecting this important part of his command, offers a parallel to those who still cling to the old-time absurdities. Ignoring the functions of the frog is not the only injury which follows the common system of shoeing. The spreading of the heel is stopped, and both of the pedal guards against concussion are nullified. That the spring which is afforded by the spreading of the "quarters" is of vast importance I am well satisfied, and also that it is nearly, if not quite, as serviceable in keeping the foot sound as anything in the mechanism of the foot, and all the contrivances to palliate the evils of a full shoe are nearly worthless.

Where the writer lived when a boy, oxen did the largest proportion of the work of the farm. It was a hilly country, heavily timbered, rocky in many places, and all the soil was nearly covered with stones. In the winter time, shoes with sharp calkins were a necessity to keep the animal from falling down on the ice-covered roads. Those which were used much in the roads in the summer needed a guard to protect the thin horn. There were two shoes on each foot, the toe portion being wide, the heel narrowed to a width of scarcely half an inch. They were fastened with very small nails, and the shoer of the ox certainly exhibited far more sense than he did when the horse was the subject. To have followed the plan pursued with the latter, he should have made one in place of two shoes for each foot, and in lieu of the narrow, thin heel, put iron enough in it to raise the pad so it could never touch the ground. He could see the importance of not interfering with the cleft in the toe; the fissures in the heel of the horse were completely overlooked, and the palpable object entirely unheeded. By putting a round shoe on the cloven foot of the ox, securely nailing it so as to completely bind both portions together, there would be a proper analogy in the systems, and the ox crippled, as well as his co-laborer.

I do not hesitate to ascribe a majority of the ailments to the effect of a wrong system of shoeing; but it is sufficient, perhaps, to consider the direct injuries. Contraction, corns, quarter-cracks, bruises affecting the sensitive portions of the foot, ossified cartilages, naviculararthritis, thrush, atrophy of the frog, etc. In the first year I spent in California I saw more corns and quarter-cracks than had come under my observation in many years previous—in fact, there were more horses troubled with cracking of the hoof than, taking them all together, I had noticed in my experience before.

The first gallop of a mile I gave two of my horses, on the Oakland course, resulted in injuries which incapacitated one entirely; the other ran many races, but with his feet so shattered that his capacity was so much curtailed that he could only display a small portion of his real powers. The first was a three-year-old filly, and in a match at Chicago, when a year younger, she easily defeated a sister to Ella Rowett, and would, doubtless, have made a fine race mare. The other was Hock-Hocking. The filly broke off a wing of the coffin bone from the concussion, consequent on the jar of the hard track and still harder shoe, and the splitting of the fibres of the horn was due to the same cause. As has been described and illustrated in a previous chapter, I invented and tested a shoe which was something of a safeguard, viz: Two plates with a stratum of rubber between. This was a step in the right direction, and while laboring to make it still more effective, by increasing the width of the heel, covering the whole space between the frog and the outside of the wall, so as to use more of the elastic substance where the whole of the concussion came, it struck me that there was a natural safeguard which would do away with the necessity of artificial appliances.

This safeguard is the highly elastic frog, when in a natural state, the spring which the commissures permit, and the dilatation and contraction in the coronary region. Taking the hoof of a colt which had never been shod, between my knees, grasping the wall with my fingers, and the thumbs pressing on the posterior part of the bars, I found that it yielded readily, and the amount of motion there was. Contrasting this with the shod foot, the difference was so great that I thought it must arise from some peculiarity in the individual, and that the colt had a more pliant wall than usual, but repeated experiments on various animals convinced me that this was one of the wise provisions to guard against injury which nature gives, and for remedying the disadvantages of the one toe in the modern *caballus* was an admirable contrivance. All other animals, with the exception of the horse, ass, zebra, quagga, etc., have divided hoofs, or padded heels, like the camel, or curved toes like the greyhound, and the guards against concussion when in rapid progression are ample. The rigid bar of iron not only fetters the motion which the commissures permit, but it elevates the frog so as to deprive it of the advantages of use, and restrains the elasticity of the upper portion of the wall.

After duly considering these facts, I felt that my elastic shoe was unnecessary, and though I had spent a great deal of time in making patterns, and had a model finished with the intention of taking out a patent, I became satisfied that there was a simple plan far more effectual. This was the use of tips, or rather a shoe which was cut off a little back of where the second nail from the toe usually comes. True, these were old appliances, and it appeared as though, if they possessed merit, they would be in more general use. Then it struck me that, while the principle was correct, the application was erroneous. The custom was to prepare the foot the same as if a full shoe was to be put on, and the tip was "feathered," running in a wedge shape from the toe to where it ended. The bars and heels were cut away so that the toe was tilted upwards, resulting in an undue strain on the tendons, while the weakened bar and thinned sole were unprotected. I make the tip of nearly a uniform thickness, a majority of them having a quarter of an inch of metal which was filed square. A shoulder was cut in the wall, and so much of the sole as the width of the web required, and all back of the shoulder, was left full and rounded with a file to protect the edge. Horses that were exercised on the track would grow more horn than was worn away, and the superfluous growth was removed when the tips were reset. The result of some of these experiments since the adoption of this system will be found in preceding chapters. The first was printed nearly four years ago (April, 1876), and the breaks were occasioned by waiting for the confirmation of the theory by the test of actual practice, or the discovery of defects which would interdict its adoption.

For the last year I have only had X X (double-cross) to test the matter fully, though there have been incidental illustrations which I have learned something from. One of these was Three Cheers. This horse also went wrong the first summer I was in California, having a "bowed tendon," which I think resulted from disease in the foot. He was of so much promise when a two-year-old that I was extremely anxious to run him, were it only in one race, being confident that he was an A No. 1 race-horse at any distance. The spring of 1878 I galloped him in tips, and he went along finely for such a length of time that I had great hopes of him "getting to a race." During the winter he had been worked on the road, with strong exercise up the slope from West Berkeley to near the University. When the track was dry enough, he was galloped on it, taking his work well, and moving through the stretches fast. Showing a trifle lame after his last run in April, I concluded to let him rest until another year, though there was very little tenderness or heat in the ailing leg. He had been in exercise from January until April. Thinking there might be some logic in the idea that when there was an injury in the back-tendon, raising the heel was beneficial, I concluded to try it on Three Cheers, and had him shod with a full shoe, with the heels a good deal thicker than the toe, in the spring of 1879. The second time he was galloped at any rate of speed, he pulled up quite lame, and the tendon was greatly enlarged. Had the gallop been severe, he would have broken down, and he did not have one-tenth of the work he accomplished safely in the tips. This satisfied me that the elevation of the heel was wrong, and in place of relief brought a greater strain on the tendons. I was led to try the shoes on him by the statement of a man who was anxious to train him, and he was so sure of the success of a peculiar application to his legs he possessed, that he made the proposition to do the training for a certain proportion of his winnings.

He took him to Pleasanton, bringing him home, according to his statement, on account of suits being brought against him for feed and board. He told me that he had given him two runs of half miles and repeat, when he made the distance easily in fifty seconds. His legs were in good shape, and as he had him shod with the full shoe, I reasoned that if he had taken the work which was claimed, there was a strong presumption of the efficacy of the shoe in such cases, corroborating the principle of elevating the heel. I became satisfied, however, that the claim was not correct, and that the horse remained idle during the period he had him.

In previous papers I have alluded to the instantaneous photographs of Muybridge, which give an accurate representation of the action in the fast gallop, and from these I am satisfied that the cause of injury to the tendons is the immense strain on the forelegs consequent on the body being carried for such a long distance while supported by only one leg. The forefoot comes to the ground at its furthest extension, and immediately after the other three are elevated, so that the whole of the weight of horse and rider is thrown on this leg. When it sustains the weight directly over it, then the pastern is so much bent that the fetlock touches the ground. With a high heel it is evident that the pastern must have a more acute angle, and the foot strikes the ground in an unnatural position. There must be such an immense strain when the leg, unaided, has to sustain the weight of the horse and the rider, and when added to these the velocity with which the animal moves, a very little additional cause may be the means of severe injury. A wrong placing of the foot, so that the other strikes it when passing, especially on the turn of the track, and surely a hampering of the natural motion of the horn, and the abrogation of the elasticity of the frog, may be a potent auxilliary in the breaking down of race-horses. Fortunately, after the body is hurled through the air, the first contact with the ground is made with one of the hind feet, then the other touches, in a twenty-two feet stride, thirty-eight inches in advance, both of these being on the ground at the same time. The first fore-foot to strike is ninety inches further along than where the second hind foot was placed, and the second fore foot fifty-two inches in the lead of the other. The stride of a race-horse, viz.: from where one of the fore feet left the ground until it strikes again—has a different movement of the legs all through. In the square trotter, such as Occident and Abe Edginton, there are identical movements twice in the same stride. The off hind foot and the near front one touch the ground nearly together, probably the hind foot a little the quickest, and that does not leave the ground until after the fore foot is elevated a few inches. Thus there are two times during the stride of the trotter when the body is unsupported, and once in the race-horse, the latter being in the air the ninety inches which marks the length of the bound. The fore and hind foot both sustain the trotter, excepting when the former is raised just previous to the last effort of he hind leg to send the body along.

CHAPTER X.

GROWTH OF THE HORN—PROTECTION TO THE FOOT.

The ill effects of rendering nugatory the natural guards against concussion are so palpable that it appears like a waste of space to reiterate the proof of the injuries that are sure to follow. But whoever will take time enough to become acquainted with the mechanism of the foot, and will give the subject some attention, can scarcely fail to see the importance of retaining the elasticity intact. Not merely that of the frog, as the spring which follows giving the quarters full play is an aid which cannot be dispensed with without serious injury. There is another thing in connection with the topic which is worthy of consideration, and that is the growth of horn. That this is much slower in the foot which has been shod, we have abundant proof, and it does not require examinations extending over a long period to become satisfied. Though the horn is deposited by ducts in the coronary band, their activity is stimulated by the natural functions being preserved. In this there is a similarity to the waste and reproduction of the muscles. Muscular effort brings a waste of the tissues, but the action that destroys also stimulates the organs which deposit the material, and the repair is accelerated so that there is an absolute increase of the motive power. Though the loss is greater, there is an extra compensation—a return of the capital with interest. There must be a judicious exercise of the muscles or there will be a diminution in the bulk, and a shrinkage which the blood has not the power to replace. In the foot there may be so much wear that the increased activity of the horn-forming vessels cannot make up the deficiency. Now, as has been demonstrated in previous articles, the foot entirely unprotected (under certain conditions) may wear away so much at the toe as to cause serious lameness. Even this is not so likely to be the case as many imagine, and in many experiments I have found that an unshod foot will stand far more attrition than was supposed to be possible. But to give the foot a fair chance, it must be in a sound condition to start with. If it has been rendered weak by a non-use of the parts which nature intended to do a share of the work, it will not stand the test under the most favorable circumstances.

The horse which has worn shoes for any length of time will soon go lame when without their protection. The smith has destroyed the capacity for resistance, and there must be a continuance of the system that has destroyed, or opportunity to recover from the treatment. The growth of the horn has not only been curtailed, but that which is deposited is of an inferior character. It may appear paradoxical to state that as the horn becomes harder it is more rapidly worn away by the attrition of the roads. It will be understood, however, when a comparison is made between the horn from an unshod foot and that which has worn iron for a length of time. The former cuts more easily when the tool used is a sharp knife, but a rasp has less effect. One is an elastic, live material, pliable as a piece of whalebone; the other a brittle substance, almost as destitute of toughness as the horn from a dead animal. Mr. Douglas, an English writer, states: "If the crust is closely examined with a microscope, its structure will be found to consist of a number of bristle-like fibers standing on end, but leaning diagonally towards the ground. From the particular longitudinal construction of the fibers, it follows that they will bear a great amount of weight so long as they are kept in a natural state. The crust so viewed resembles a number of small tubes, bound together by a hardened, glue-like substance."

"Whoever has seen a Mitrailleuse gun, with its numerous barrels all soldered together, can form a very good idea of the peculiar structure of the crust (or wall), especially if they were likewise to imagine the tubes filled with a thick fluid, the use of which is to nourish and preserve them." Before microscopic observations revealed the existence of these minute tubes, it was thought that the horn lay in layers like the leaves of a book, and the tubes are so arranged as to form these thin strata. The layers are readily separated, and it does not require a long soaking to make the division. The agglutination is more rigid between the tubes at right angles to the layers, and consequently a great force is necessary to rend them asunder. But if the tubes are hardened from dryness, the diameter must be decreased, and there cannot be a full supply of the fluid, the duty of which is to keep the horn in proper condition. The atrophy of the frog in the shod foot is a striking testimony that the nourishment is lacking. There is no longer the stimulus which exercise gives, and though there is no waste from wear, the supply is cut off. The preventing the wear at the toe by the application of a tip—or, more properly speaking, a lunette shoe extending as far back as the point of the frog, or a trifle further back—does not seem to retard the growth of the horn at the part covered by the iron. This probably arises from the posterior portion of the foot being so much more pliant, and the stimulus to the secretory vessels of the coronet sufficient to keep up the healthy action of the whole of the circle. Since using tips I have noticed that the toe would appear long before they had, apparently, been on a sufficient time to account for the extra growth. It was ascribed to the wearing away of the heel, while the toe was guarded; but a late experiment shows that this was only partially the cause. On the 19th of January, X X was shod with tips weighing seven ounces each. Previous to that he had been without anything on his feet, and was turned, during the day, into a small lot. There was very little wearing of the horn, and when the tips were put on the heels had to be lowered to give the proper bearing. When this reduction was made, a gauge was set so as to give an accurate measurement from the ground surface to the junction of the hair and horn. At this date, the 16th of February, four weeks since the tips were put on, the toe is quite long, altogether too much so to give a proper bearing to the foot, but the gauge shows that the wear at the heel has not been quite as much as the growth. He has been used, with few exceptions, daily on the road, and as most of the time the streets were wet, there was more wear than there would have been in dry weather. The streets and drives in Oakland are formed of macadam, with gravel or finely-broken stone on the surface, and the horn wears away more rapidly than in the summer. This experiment was further proof of the necessity for bedding the tip, as heretofore described, into the foot, and also that it is important to frequently reset them. The growth of the horn, between the tip and the foot, being more rapid than when the full shoe is used, the iron must be removed whenever the bearing is thrown out of the proper level. As steel tips, hardened, are used, there is little wear in the metal, and if the heel is not as low as it was, there is the difference to overcome. The best plan is, to sink the tip rather more than to bring it to the level of the heel, and the action of some horses is benefited thereby. But as the same nail-holes can be used for two shoeings, when steel is the material, and not longer than three weeks intervene between the "removals," there is no injury from extra perforations. The fewer holes in the horn, the fewer tubes will be cut, and the supply of fluid they carry only slightly interfered with.

The objection to the use of tips which is the most frequently and persistently urged by those who denounce the practice, is the claim that the heels will be bruised if left as nature made them. This has a plausible look, and without the person who hears it offered has made some progress in acquiring the art of forming his opinions on what can be proved, is very likely to carry conviction. In former papers the fallacy of the claim was shown, but as such wide intervals have elapsed since the publication of the series, it becomes necessary to reiterate. With the drawbacks arising from a want of connection, and the necessity of going over the same ground, there are advantages in being able to make more positive statements resulting from later experiments, and even if the repetition may be something like a tale many times told, the subject is of importance enough to warrant the rehearsal. The importance of taking care of the feet of the horse has always been realized, and there have been so many plans for counteracting the ailment which is most prevalent that it appears singular that so little progress is made. The trouble has been that a large majority of those who have been given advice have been themselves misled by the idea that protection to all parts of the foot was indispensable. The few who have advocated the natural method, so far as domestication would permit, have been forced to meet this obstacle, aggravated with the tendency of horse owners to let others think for them, and an unwillingness to give a fair trial to so simple a remedy. When the groom and the smith told the owner that bruises of the sole would surely result if there was not an iron barrier to protect, it appeared reasonable, as it may have been that when a boy he had endured the pain of "stone bruises" from running barefooted, or "stubbed his toe" on a projecting rock. The remembrance of sitting in agony, rubbing the foot in a grip as hard as he could make his fingers clinch it, swaying his body and gritting his teeth as some relief, gave emphasis to the claim, and though he may have gone to the shop, firm in the determination to try the method he had read of, the words had weight and his good intentions were overcome; overcome by dogmatic opinions without argument or logical reasoning to sustain them.

If he remembered as well the acute pain caused by a small pebble or even a kernel of corn getting between the foot and the shoe, he would know that there were other casualties beside bruises to guard against. But there is nothing analogous between the foot of the biped and quadruped, and the stone bruise on the barefooted boy, and that which causes corns in the horse, are widely different. No matter how thick the skin on the heel may have become, it is a slight protection in comparison to the walls, bars and sole of the horse. That is, when all parts of the foot of the horse are in a natural condition. When the smith has pared away the natural defenses so that it will "yield to a strong pressure of the thumb," as is recommended in the essay which drew the $500 prize in England, forty or fifty years ago, the only plan is to raise it so that it will not perform the duty it was intended it should. A little more paring and the sole would be entirely cut away, for when it will give to the pressure spoken of there is a very thin layer of horn left.

I have astonished quite a number of visitors by showing them the results of wearing tips in the strengthening of the posterior portion of the foot. The angle between the bars and the walls, the seat of corns, is filled with a dense, elastic material, which does not show a trace of the discoloration which is nearly universal in shod horses. The frog is as elastic as a piece of solid rubber, and the point of it extends to within a short distance of the toe. A few days ago, I was resetting the tips on X X and fortunately a neighbor was watching the operation, who adhered firmly to the old, old methods of shoeing. He had frequently urged the necessity of protection, and could not see why the quarters and heels should require it less than the toe. It was a good opportunity to make one convert, and without acquainting him of my intention, I led the conversation to shoeing. He combatted the statements energetically, laying the greatest stress on the liability of bruises. He admitted that the chief amount of friction was at the toe, and an old shoe which had the toe entirely worn through at the outer edge, with quite a thickness of iron at the heel, was proof of that fact if he had not admitted it. The blow came on the heel first, and as the foot rolled over the pivotal point, the friction wore away the metal.

"But," he said, "the fact that the blow is mainly on the heel shows the necessity for protecting that with a hard material. In a natural state horses roam over an uncultivated country. There is either growing herbage, fallen leaves, or snow, to make a cushion for the foot to strike upon; but on traveled roads, on the paved streets of a city, on the solid body of macadam such as the streets of Oakland, there must be protection." I replied with the query: "Mr. J——, what, in your opinion, would be the effect of driving a horse wearing tips, for years nearly every day, and fast enough to be called a good traveler, on these Oakland streets, and on the roads in the vicinity?" "Bruises on the heel, and the wearing away of the unprotected parts so as to make the animal too sore for use, and if persisted in there would be violent inflammatory action which ultimately would destroy the foot." "Very well," was the answer, "and if you have an hour or two to spare I will endeavor to convince you of the error of your views." This conversation took place in the house, and before going to the barn I showed him how the horse had been shod for the past six years, by reading extracts from journals. For six years he had worn tips, with a few exceptions. Twice full shoes had been put on in front, and at other times three-quarter shoes, but with the two exceptions the inside quarter had been bare. The three-quarter shoes had been put on for the purpose of testing the effect on the action, and not for any necessity for protection. From August 17, 1880, until October 4, 1881, he was driven without either tips or shoes on his hind feet, and it was rare that there was a day during the time when he was not driven. The tips which he had on were set the 19th of January; and, as was stated in the previous article, there had been so much growth of the horn that the toes were too long. The guage showed that there had been a trifle more growth at the heel than the wear. On the 16th of February the tips were pulled off and the extra horn pared away. There was a slight discoloration immediately under the iron, but this was very shallow, and my friend remarked that the hoof was in capital shape; and, doubtless, I had some kind of hoof ointment which was the means of keeping it in such fine condition. I assured him that the day of all these nostrums had passed away, so far as I was concerned.

Soaking-tubs were discarded, and only at rare intervals was a wet sponge applied to the hoof. The place for the reception of the tip was cut, leaving the shoulder which the end of it was to rest against nearly half an inch in depth. It was a work of some time to properly prepare it for the tip, and it is a job which always "starts the sweat" in streams. The trouble is to get a true bearing, and this is far more difficult than when a full shoe is to be put on. In this latter case, the eye tells at a glance if it is "out of wind," whereas the shoulders are in the way, and the plan followed is to make the foot surface of the tip as true as I can file it, and then make the horn comport with the tip. Of course, the outside of the tip is fitted to the wall before the surface-filing is done. When the tip was nailed on, the shoulder was an eighth of an inch above the ground surface of the tip when the foot was held between the knees, and this had to be cut away to give the proper bearing. "Now, Mr. J——," I remarked, "you will see a foot which has travelled over these Oakland streets for six years, besides galloping with the trotters, trained and running races in harness, worn toe-weights, side-weights, undergone all kinds of experiments, hacked about to all sorts of vehicles, and with occasional long drives and quite rapid ones. According to your ideas the foot should be ruined beyond hope of redemption. You acknowledge that the outward form is nearly perfect; it would be entirely so if he had never worn shoes; and now we will look for the bruises which a short time ago you said were sure to be found." It cut readily, and the knife sliced off a piece from the shoulder to the heel and about a sixteenth of an inch thick. It was from the inside quarter, and beneath it the horn did not show a particle of discoloration, as had been the case at the toe. We handed it to him, and he bent and twisted it, put it between his teeth to test the toughness, noted the width of the frog and the india-rubber-like elasticity it displayed. After a minute and careful examination he admitted that his previous opinions were certainly erroneous, but before he made many remarks I interrupted him, with a request that he should pay me another visit in a week from that time, when a new set of tips were to be put on Anteeo, as I desired to show him a perfect foot. That of X X had been injured by wearing shoes, and came very near breaking down on that account, while the colt has never worn shoes in front.

CHAPTER XI.

A NATURAL FOOT VS. A PERFECT FOOT.

A natural foot may not be perfect, and then there may be advantages in artificial appliances. But in a large majority of animals the feet are nearly right, if Nature has not been tampered with in the effort to make them so. If permitted to exercise on the same kind of ground which would be chosen in a natural range, it would be rare, indeed, when there was much divergence from the true form. By a natural range we mean that part of a country which would be selected by horses which could follow what is called instinct, but which is so nearly allied to reason that we never could see where the difference was. With the herbage equal, there will be a general resort to the highest ground, and if it is necessary to leave it to find water, as soon as thirst is assuaged the band will return. Left to themselves, the firm ground is invariably chosen, and, though they may be driven to seek refuge from flies by standing immersed as much as possible, there is a disinclination to wet their feet and legs. The whole system of soaking the feet, stuffing with clay, cow-dung and hoof-pads, is entirely contrary to nature, and the use of hoof ointments of any description a departure which will end in injury. Even the washing of the legs is carried to an extent which does damage, and, as a rule, water should never be applied to the horn. It may require more work to cleanse the feet of the dirt which has been gathered in a drive through the mud; and to wash them and the legs with water is a labor-saving contrivance for the groom, but is a positive injury to the feet, and has also a tendency to crack the heels. There is trouble enough from the liability of the skin to crack between the foot and pastern, without aggravating a tendency which is unavoidable at times, with the best of care. Cracked heels are one of the most pestering things which the trainer has to contend against in the category of minor troubles, and frequently they are of major importance. Not alone from the pain which causes the animal to shorten its stride, but, in aggravated cases, the inflammatory action extends to the tendons, and the breaking down which results, if the work is continued, is due, perhaps, to the use of water. The erroneous impression which is so prevalent that moisture is beneficial to the foot, has done great injury.

This has arisen from the palliation of the injuries which resulted from improper shoeing, and from the mutilations of the foot, which it is thought that it is the imperative duty of the blacksmith to accomplish. This renders the horn brittle, and temporary elasticity is gained by the artificial moisture. The natural foot has none of its functions impaired. The secretory vessels are allowed full play, strengthened by action, and the deposit, consequently, is of the right character. The thousands of minute tubes are filled with healthy fluid, and these are not marred by knife or nail. When worn away by the attrition between the foot and the ground, there is the same safeguard against depletion as follows the searing of an artery, and the lower portion is glazed with the friction, which also accomplishes another beneficial duty, viz.: the prevention of absorption of moisture. Anyone who will take the trouble to examine the natural foot of the horse will find that ample provision has been made to exclude water. The horn is coated with an enamel which is as impervious to the entrance of moisture as a plate of glass, and the sole in a natural state will also exclude it. This phase of the foot question was forcibly presented from a conversation with a gentleman a few days ago. He had formerly been a blacksmith, but for the past number of years has been engaged in driving cattle, formerly from Texas, and later from the mountain ranges. We obtained much valuable information from him, which will be briefly, and only in small part, alluded to now, as we hope to have a longer interview, and which we will endeavor to make a more satisfactory record of. He has had an experience which falls to the lot of few, and possessing keen powers of observation, he has been benefited by the experience in an unusual degree. With 250 saddle-horses in his employment, engaged in as arduous duties to try the feet and legs as could be imagined, we were prepared to hear him staunch in his advocacy of a natural foot, though we expected that more of the antiquated notions of the shoeing-forge would be exhibited. One remark he made will be the limit in this article. "Give him," he said, "a horse with bad feet, no matter how bad, so long as the natural functions have not been entirely destroyed, and by the time there has been growth enough to remove the brittleness from the nail-holes down, the cure in most cases will be accomplished." His treatment was to turn him barefooted on a proper range, and in place of this being the marshy ground which is usually recommended, it was hilly, rolling land. This he further illustrated by experience with a large band of cattle which he bought. They had become so foot-sore that they could not be driven, and were then feeding on the low ground. After the purchase he transferred them to the hills, and in a short time they were able to travel. The brittleness below the nail-holes is doubtless caused by cutting the tubes. The perforations which are made by the nails cut off the supply from above, and the cutting to prepare the horn for the reception of the shoe empties the lower part of the tubes of the fluid which gives vitality. We have had a capital illustration of this in Lady Viva, a filly which was two years old on the 3d inst. Nearly a year ago she suddenly exhibited lameness, and most careful examination failed to discover the cause of it. Feeling that it was probably in the foot, every method was followed which promised elucidation, but without avail. It was several months before the cause of the trouble was manifest. She had a habit of climbing on the fence, resting her feet on the bar to which the boards were nailed, in order to look over, and, doubtless, got a redwood sliver in her toe, at the juncture of the sole and wall, which broke off so far in the horn as to escape the search. This came through at the coronet, leaving a crevice of three-quarters of an inch. This of course caused a separation of the horn, and though, after there was growth enough to relieve the pain, the lameness vanished, it had been of such long standing as to interfere with the proportion of the foot. The one affected is still a trifle smaller than the other, and with higher heels. This came from the inflammatory action while the foreign substance was in the foot, and by keeping that part cut down so that the frog was brought into play, it has spread to nearly the natural proportion.

The crack in the horn was nearly three quarters of an inch above the ground surface of the toe, when we began to drive her on the road, and not very long afterwards, the toe broke away, leaving a gap the width of the crack. This compelled putting a tip on that foot, and in order that there should be no disparity, a tip was also put on the other. The tips weighed nearly four ounces each, and were put on the 15th of February. The driving commenced on the 8th, and she was driven entirely on the macadamized streets of Oakland, the others being too muddy to exercise her upon. The well foot was slightly worn at the toe, though it did not require any "protection," and the intention was not to use even the tips for some time, if at all. On the 7th of March she lost the tip from the broken foot, and returned from the drive with the toe broken for half an inch or more on each side of the gap. That this break came from the brittleness caused by the separation of the tubes is evident, for the hind feet, which have not been protected, have not broken a particle, and are in capital condition. On the following day a tip was put on, and in order to set it properly, the horn had to be cut sufficiently to reach that which had been deposited since the break at the coronet. There was a palpable difference between that and the brittle portion, the former being much the toughest. The junction of the wall and sole gave evidence of the injury, there being an entirely different appearance at the point of contact from that of the well foot. If the examination had been thorough enough at the time of the injury to discover the sliver, the extraction of it would have saved the foot; and though the cutting necessary might have resulted in more acute lameness, it would have been for a brief period. We were much disappointed in not being able to try the experiment of using a natural foot without protection of any kind, but in this case will have to confine the tests to the hind feet. There is more of a sliding motion behind than in front, and as a general rule, the hind shoes have to be replaced while those in front are comparatively unworn. Some of the fast trotters will wear a hind shoe entirely out in two weeks, the front having only a slight bevel at the toe.

CHAPTER XI—Continued.

Reason for Want of Correction—History of Anteeo.

It may appear somewhat egotistical for the writer to illustrate the lessons on shoeing with histories of his own horses, but owing to the absence of other data, that course is forced upon him, and without entering into minute, and, perhaps, in some cases, tedious descriptions, the explanations would not be clear. The few which were available for experiments have necessarily extended the time to a long period, and it is now six years since the departure was made. In every case, excepting the one which will be the subject of this chapter, the animals experimented with had worn full shoes, and though this gave the opportunity of contrasting the systems, it is evident that others were needed to show the benefits which would follow from the functions of the foot never having been interfered with. The colt, Anteeo, gave this opportunity, and so far as can be determined by a trial extending from the time he was nearly fourteen months old until now, when thirty-four lunar revolutions have elapsed since the date of his birth. The history of these twenty months of childhood had to be quite full in order to give, with clearness, the state of the case, and, in order to show our friend the actual every-day life, without the trouble of going over the pages of the journal, the information was condensed into the following synopsis:

Anteeo was foaled on the 5th day of May, 1879. This being rather a "late colt" for California, we determined not to breed the mare, and permitted him to suck as long as it was deemed beneficial, though after he was six months old, it would have been as well to wean him, and the two months more were of little benefit. Before he was weaned he was led a few times by the side of the mother, showing fine action when trotting. The day he was eleven months old, led him by the side of X X, and occasionally afterwards; and on the 26th of April, timed him a quarter of a mile in 1:10. . This was certainly a fair beginning, and on the 29th he made 1:07. Nothing more was done with him until June 3d, when, led by the side of X X, he trotted a quarter in 1:05; and on the 5th and 7th in 1:04. On June 12th, he was harnessed and driven by the side of X X—the horse in the shafts of a breaking-cart, the colt on the off side, his traces hitched to a whiffle-tree, which was fastened to a bar tied to the shafts, and projecting far enough for the purpose. He had been driven previously with the harness on to accustom him to being reined, etc., and drove very quietly. On the 22d of June, he was shod with tips, weighing about four ounces each. His exercise was confined to leading him by the side of X X, and on the 26th of June he trotted the "back-quarter" in 58 seconds, and the homestretch in 56 seconds. This was a manifest improvement, which was ascribed to the tips favoring his action; and on the 14th of July, he was hitched to a breaking-cart, and he went quietly. On the 20th of July, he was harnessed to a light sulky, but he could not trot within 15 seconds to the quarter as fast in that as when led. This was thought to be caused by striking his hind foot when going under the front, and rolls on his hind pasterns improved him, but not enough to show the speed when he was led. To test the effect of the toe-weights, on the 6th of August, tips, with a "spur" welded to them, were put on, spur and tip weighing seven and one-half ounces, and on the following day he was driven with them and toe-weights of three and one-half ounces. He trotted better, but pulled up quite lame, which we thought was occasioned by paring the foot. The boy who held him at the shop ascribed it to a twitch the blacksmith gave his leg. He had driven the nails and taken the foot forward to clinch them, when the colt tried to get his foot away. The smith had a grasp with his left hand immediately below the knee, and the foot in his right, and when the colt struggled he snatched it violently upwards. The diagnosis of the boy proved correct, as ultimately there was swelling just below the trapezium. Thinking it more trivial than it afterwards proved, on the 9th a pair of hind shoes were put on, in order to hold a scalping-boot, but the swelling and lameness continuing, notwithstanding fomentations and cooling appliances, the shoes were pulled off, a light blister applied, and he was allowed to run in a small lot during the day. October 3d he was taken up and driven barefooted a quarter in 58 seconds, and on the 10th in 54 seconds. Previous to this we had given up all expectation of trotting him in the Embryo Stake of 1880, though the move was so satisfactory that it was concluded to give him another trial. On the 13th of October he was shod with tips in front, weighing six ounces each, and shoes behind, so as to attach scalping-boots. On that day he was weighed, and for a colt of seventeen months old he showed plenty of bulk, as the record was 810 pounds, and $14\frac{1}{2}$ hands high. The 23d of October he was a trifle lame, and his work was limited to a jogging, with an occasional spurt of speed. On the 28th the tips were reset and lighter hind shoes put on. The tips had been worn some, and were filed down to $4\frac{1}{2}$ ounces each, including the nails, and the shoes the same weight. He was driven the following day in 3:48, and we were satisfied that he could trot the mile better than 3:20. This view proved to be correct, as on November 5th he won the stake in $3:17\frac{1}{2}$, the last half in $1:37\frac{1}{2}$, and evidently could have gone considerably faster. He was worked slow, and on November 10th the tips were reset, and three-quarter shoes, covering the toe and outside wall of the hind feet, put on. The use of the three-quarter shoes has been partially explained in former articles, but later experiments have shaken the old belief, and the effects are so different from what was formerly anticipated, that it is likely there will be a complete change. It will require one or more chapters to present the features, and as this is mainly to show the result of the use of tips, the consideration will be postponed. On the 12th of November he was driven in 46 seconds, and this was the fastest time he made in his yearling form. He was driven a few times on the road and ran in a small lot during the winter, when the days were suitable. On the 4th of April, 1881, his feet were trimmed, and he was led to the track. On the 15th he was hitched to a breaking-cart, driven a few times, and again turned out.

In explanation of the erratic manner he was exercised during the summer of 1881, a brief statement is required. A colored boy who accompanied us from Chicago was the only help kept. He broke the colt and worked him and drove him in the race. From a fall, when hoisting hay into the barn, supplemented with a cold taken when on a duck-hunting expedition, his lungs were affected, and we did not feel that he ought to be permitted to risk the labor attending working and taking care of a colt none the easiest to manage. He was extremely sensitive about any one else taking care of him or others to drive, and we were too much occupied with work which had to be done to give him personal attention. Thus on the 5th of May he was driven to a breaking-cart and turned in a small lot as before until June 1st, when he was driven slow in breaking cart until the 9th. On that date put on tips, weighing four ounces each; the day before he was shod behind with shoes covering the toe and outside, leaving the inner bare. After putting on the tips he was driven to a sulky for the first time since November, and moved fast. He was driven occasionally until June 29th, when having to attend the races at Sacramento he was turned in lot. The 11th of July put on tips weighing five and one-half ounces each, and he wore them until August 8th when they were replaced with others of seven and one-half ounces each. While wearing those he trotted a quarter in 43 seconds, and on the 29th of August they were taken off and others put on of three and three-fourth ounces. With the light ones he handled himself more satisfactorily, though having to attend the fairs of the Golden Gate and State Agricultural Society he was turned out, the tips being always left on when the time was expected to be short. He ran in the small lot during the day from the first to the 28th of September. On October 6th put on weights weighing five and one-half ounces each, and on the 14th drove him a mile in $2:54\frac{1}{2}$, the last half in 1:24, and keeping him all the time well within his rate. On the 18th he trotted a quarter in 40 seconds, and we felt assured that he could show a "thirty gait" if called upon. We drove him in this summer's work, though, as is evident, it was of too desultory a character to be called training. Still it was a fair test of the tips, both as to the effect on the feet and to improvements in trotting. We were laid up part of the time, confined to our room from the 24th of October, and from then to the race he was in the hands of another colored boy who came from the East with us.

He won the second Embryo with him, making the third heat in 2:52, when again he was turned out. On Jan. 2nd of this year tips were put on, the same which he wore in the race, and from that time until the 28th he was allowed to run in the lot or exercise by leading when he was harnessed to the breaking-cart. He was driven on the streets of Oakland, daily, generally long drives, and at times fast, the tips having been reset on February 7th, but merely paring enough so that the old nail-holes could be used, and on the 23d of February was the day fixed for our friend to witness the operation, and see the condition of the foot. The tips were semicircles with a diameter of five inches, and this brought the posterior portion about three-quarters of an inch in the rear of the point of the frog. They weighed seven ounces each, and were three-eighths of an inch in thickness at the end, being a trifle thicker at the toe, the width at the toe seven-eighths of an inch, and at the ends five-eighths. They were made of tool steel, the nail-holes countersunk on the ground surface, in place of being "creased," the countersinking deep enough to imbed the head of the nail. From so little horn having been cut away when the tips were reset, there was a superabundance of horn, and the shoulder for the end to rest against was cut deeper than was required to imbed the tip. The rasp was used to lower the outside of the wall, beveling it to the gauge-mark on the outside, and back as far as the shoulder. We then called the attention of our friend to the heels of the feet, and width and elasticity of the frogs. There had not been wear enough to reduce the wall to the level of the sole and bars, and it was prominent and without a break. From the shoulder for the tip to rest against to the junction of the wall and bars, was a little over two inches, and it would be impossible to improve on the general contour of the foot. This our friend admitted, and when the knife was used to cut away the superfluous horn, back of the rasping, there was only a slight discoloration, which the first shaving entirely removed. The enamel having been cut with the rasp, it was an easy thing to cut the other portion with the keen, thin hoof-knife, though the thinnest shavings were as tough as the best quality of whalebone. It was a work of time to get a smooth, perfect bearing for the tip to rest upon, and when a fine rasp and file was brought into play to perfect the fit, it cut slowly. There were three nail-holes on each side of the tip, and our friend thought two would be sufficient. The explanation for using three nails in place of two was that we desired a guard against any chance for motion between the end of the tip and the foot, and that where so long a tip was used the rear nail was at the end. But we are now satisfied that a semicircle extends too far back, and that one-third will be ample protection. We have found that in a great majority of instances the rear nail, on one side or other, was broken, and this was caused by the expansion of the foot, and reaching as far forward as the nailing. The nails used being next to the smallest size, the constant pressure from giving the foot an opportunity for free expansion broke the nail; whereas, if limited to one-third of the circle, that would not be the case. That this would be ample protection we have not the least doubt; and then three nails, one at the toe and one at each end, would hold it firmly in place. At all events, when the foot is in a shape which will permit a trial, I shall make the experiment. My friend was also pleased with the system of nailing, viz., driving the nail from the inside of the wall through it, in place of splitting the layers; and also at the use of a small gouge to remove only enough of the horn to receive the "clench," in place of cutting the enamel from one nail to the other, as when the file is used. The two exhibitions removed all doubts, and he was lavish in his praises of a plan which had left a natural and perfect foot.

CHAPTER XII.

HISTORY OF ANTEEO CONTINUED—SKETCH OF ANTEVOLO.

The history of Anteeo was carried in the preceding chapter until February 23, 1882, when within seventy days of three years old, and a *resumé* to the present date, February 1, 1883, will show how the tips have answered in his case. It would be tiresome to enter into the history as minutely as I could give it, as everything in relation to his shoeing and work has been noted in the daily journal.

His case has probably elicited more discussion, or, rather, more adverse comments on the method he was shod than will ever occur again. He was brought prominently before the public from being engaged in some important stakes, and all his shortcomings charged to the tips. Had it been otherwise than that I was thoroughly imbued, and obstinately confident of the correctness of the principle, I would have surrendered to the universal clamor. Having the "courage of conviction" I was not to be moved by appeals, ridicule, or the conjectures of people, many of them being as ignorant of what they talked so learnedly about as if they had never seen a horse. Others had strong arguments, that is, when the guide is previous opinions. Not one of them had tested the difference between shoes and tips, and consequently their reasoning was purely theoretical. It is also true that on my part I could not say authoritatively that Anteeo would not trot faster with full shoes, as that was something he had never worn on his fore feet; but from all those that I had tried the change upon trotting faster with tips than when wearing shoes, the inference was that he would not be an exception. The objectors who presented the most logical reasoning based their arguments upon his action.

He has very little "action," when that term is used to express bending of the knee and hock. He is rather a "short strider," and when going at 3:30 gait is prone to hitch, sidle about, swing from one side to another, and to a person who only saw him when jogging he would convey the impression that he was unlikely to make a fast trotter. When going fast he moves as squarely as it is possible for a horse to trot; and the only thing I would care to remedy is the shortness of stride. It certainly appeared reasonable to expect that more weight on the front feet would remedy this, and it also seemed that toe-weights would be beneficial. I tried weights on different occasions, and he would not trot so well, excepting in one instance which will be given hereafter. In order to fully understand his case it will be necessary for me to recite other peculiarities. He was foaled where I reside, in Oakland, and I have the use of a lot of about an acre. He and his dam occupied the lot without other company. The colored boy alluded to before was continually petting him, and so much did he think of the colt, that he would permit him to bite and play with him without correction. Before he was weaned it was unsafe to go into the lot without a whip or stick to keep him off. It was not much trouble to break him to harness, though from the first he was stubborn, and severe punishment made him more determined in his obstinacy.

This was partly inherited, partly the result of familiarity between the boy and Anteeo when a foal. The inheritance came from Bonnie Scotland, who the English writers say was the most sluggish horse in his exercise ever trained in England. I bred and owned two colts by Bonnie Scotland which had the same disposition. They were brothers, and the elder was completely spoiled by severity; the younger, who was treated with invariable kindness, outgrew the obstinacy and became free and pleasant. The elder was one of the fastest horses I ever saw, running a quarter of a mile in twenty-three seconds in his training shoes and with his weight up, and apparently could go any distance. Severe punishment resulted in utter worthlessness, either to run or drive, as when broken to harness it was impossible to drive him away from home at any other pace than a slow walk, although when turned to come back he would trot at a three-minute clip. If whipped going away from home he would stop. If given a sharp blow as he was coming back he would kick with terrific vengeance. From that experience I knew that it would not answer to punish Anteeo; and though there was a perfect deluge of advice, the general purport of which was severity to the pitch of cruelty, I treated it the same as that to replace the tips with shoes, and kept my own course. That this has been correct is apparent, as in the last few weeks he has taken an inclination to go, and I have the fullest confidence that hereafter there will be no more trouble with him on that score.

It is necessary to be made acquainted with this peculiarity of temperament in order to understand fully why he would show at times a flight of speed, and then in his races and at other periods not trot within ten seconds as fast. That presented in so brief a manner, I will take up the discourse from the time Mr. J. watched the setting of the tips. He was not driven from early in November, 1881, to January 28, 1882, and then only occasionally to a heavy breaking-cart, generally on the road. Until May 25th he was driven to a lighter cart, part of the time on the track, with fast work once in a while. The lighter cart will probably weigh 130 pounds or more, having elliptic springs, and being strong enough to carry two men. On the 25th of May he was hitched to a sulky the first time since he trotted in the Embryo the November before. The tips had varied in weight from three ounces, those that were worn by use, to seven ounces, the heaviest I had made; usually five ounces or six were the weight of those used.

In the meantime I tried a different shoe on the hind foot, a description of which will be given in the Appendix.

On the 31st of May I commenced galloping X X with him, in order to encourage him to go along without so much urging, and from that time his fast work was in company with the galloper. June 7th commenced working him "two-and-two," finding that he was more inclined to trot the second mile than the first. This kind of work was kept up, and on June 15th, "on the repeat," he trotted the two miles in 5:20—the last mile in 2:38. On June 27th gave him three heats of two miles, the last mile of each being 2:40, 2:39¼, 2:38¼. The first mile of each heat I drove him as fast as I could, but 2:41 was the fastest, and that in the third heat. Being so busy on account of the work incidental to the publication of *The Breeder and Sportsman*, I did not drive him again until the 17th of July, his only work being jogging to the cart by the man who took care of him. July 27th I put new tips on him, weighing six ounces each, and on the 29th is the following entry in the journal: "Jogged Anteeo three miles to cart, hitched him to sulky and then gave him three miles at a good pace, moving in places fast. Scraped and repeated him two miles in 5:25, moving through two stretches in 37 seconds each. After the work he played coming home." On the 1st of August he finished a strong drive of three miles by trotting the home stretch in 37 seconds, which proved that the work three days before had not been detrimental. On the 3d of August he was attacked with the "pinkeye." This disease practically laid him up until September 16th, although he was worked occasionally, which I am satisfied was an error. The sickness prevented him trotting in the Occident Stake at the State Fair. On the 26th of September we resumed the two-mile work, endeavoring to prepare him for the Stanford Stake. The swelling between his jaws suppurated and broke a few days before that race was trotted.

In the race, when "warming up," after going at a good pace, he trotted half a mile in 1:16½, and a quarter in 36½ seconds, and yet he was beaten in 2:34½, 2:36½, 2:40, 2:38. The cause of this I will endeavor to explain after a few more illustrations. The Stanford Stake was trotted the 21st of October, on the Bay District Course; on the 25th he was brought home, and on the Oakland track I gave him three heats of two miles, in the last of which he trotted the second mile in 2:36½.

On the 30th I drove him three heats of a mile as fast as I could, scraping him between the heats. After this I gave him a heat of two miles, and he trotted the last mile in 2:30¼. The fastest first mile I could drive him up to this time was 2:39½. He made a poor showing in the Embryo Stake, but on November 26 the following is the record in the journal: "In the afternoon I drove Anteeo. Jogged to track and two miles on it to the cart. Hitched him to the sulky, X X galloping to another. Went two rounds of the track the reverse way, the last at a good rate, then turned and gave him a mile in 2:45. Slight scrape and repeated him in 2:41. Again scraped lightly, and after scoring a few times, drove him a mile, with a break soon after starting, in 2:35. Keeping on, the timers—James Garland, George Palmer and Johnson—timed from the quarter-pole, the last three-quarters in 1:54; and as the first quarter was as fast as the others, that mile was probably made in 2:32." On December 5th he trotted the last mile of the third heat in 2:34¼. One more illustration will be sufficient. After strong work from the 4th until the 12th, I determined to again try him with toe-weights; though on previous occasions he did not trot as well. On the 21st of November three-quarter shoes were put on his hind feet, at that time weighing seven ounces, and on the 4th of December I put on tips of five ounces each. By the 12th the hind three-quarter shoes would not weigh over five ounces, as when pulled off a short time afterwards they were reduced to three and a half ounces. He was thus rigged: Ordinary walking or ankle boots all around. Quarter boots, weighing four and three-fourth ounces each, on his fore feet, and toe-weights of three and three-fourth ounces each. Thus there were about thirteen ounces on each fore foot, allowing for the eight days' wear in the tip. Contrary to the previous custom, he was driven alone. Walked to the track, nearly one mile, harnessed to the sulky, jogged two miles, the reverse way, moved up the homestretch, around the turn and half way down the homestretch. When turned he seemed to want to trot; none of his usual stubbornness or mean actions. He made the first quarter in 38½ seconds, went to the half mile in 1:15¼ and the mile in 2:31½. I never drove him a mile so easily, never touching him with the whip or moving the bit in his mouth. On the 8th I drove him three heats, the slowest in 2:42; and on the 9th he was worked sharply for three miles, and trotted a heat against Bonnie and Fred Arnold.

The opponents of tips have laid great stress on the trotting of

Anteeo, and I have given this long, and, it may be considered, tedious account, in order that there may be a proper understanding of the case. In the first place, for a colt foaled May 5th, 1879, and with so disjointed a schooling as the record shows, it may be considered a very fair rate of speed. To take the maximum, 2:30¼, and it is more than a fair showing for a three-year-old, if even the education had been better managed. Had the tips been so detrimental as is claimed, so diametrically opposed to speed, there never would have been a display, and slow time would have been shown at every trial.

There could not be soreness, as the harder Anteeo worked the faster he trotted. Intervals of rest invariably added many seconds to the mile, and fast quarters, halves and miles were made at the finish of long heats. And now for the explanation of the erratic exhibition, and the causes why the second mile would be faster than the first, and the greater the number of the heats the greater the increase in the speed. When fresh he wanted to rebel. In that situation he was determined to resist the attempts to urge him to more rapid movements, and anything like severe castigation resulted in a still stronger will to thwart the driver. When he became wearied—not so tired as to prevent him from trotting—the obstinacy gave way, and then he was willing to go along. When he trotted the half mile in 1:16½ at the Bay District he had been driven two miles as fast as I could work him along, and if it had been permissible to move once around the track, and then get the word, he would have made a good performance. The mile in 2:31½ on the 12th of December, so contrary to all his previous actions, I ascribe to the weights distracting his attention at that time, when in the prior failures the same effect did not follow. Then there is a manifest improvement in his disposition, of which that may have been the forerunner. He has become more docile in the stable; or, more properly, not so mischievous. Heretofore he seemed to have an idea that men were playmates; that it was all right to jump on them, give them a nip or run against them, just as a colt is likely to play with another. Until the last few months he could not be led in hand without constant threatening him with a whip, and so I fixed a strong piece of bamboo with a snap in the end of it to keep him off. When in his box he had the same inclination—not a particle vicious, as his countenance would indicate good nature, but ready for a frolic at every opportunity. The bamboo had to be used to snap into his halter-ring, and the halter I fixed so that the bridle could be put on and the halter removed afterwards. This was effected by a buckle on the noseband; and it also gave the opportunity to replace it before the bridle was removed. Without this precaution he would try to catch the person bridling him by the leg, exactly similar to the action of a colt when playing with another. This mischievousness doubtless came from the plays with the boy when a suckling, and the endeavor to remedy it by such severity as people advised would have resulted in confirmed vice.

At the present time he is as sedate as need be. Comes to the call of his attendant in the stable, and will walk as decorously as a quiet mare. That there is a change as well in his disposition as in other respects, is apparent, and I feel quite confident that he is a different animal in still more important respects. I drove him December 15th, and from that time he has been exercised on the road by his groom. On the 18th of this month—January, 1883—I put tips on him weighing three ounces each, his hind feet bare. On the 21st, 22d and 23d I drove him to the track in the light cart, having put the three and three-fourth ounce toe-weights and quarter-boots on him, and for the first time in his history I could move him through the stretch after jogging him a couple of miles. He not only would go through the stretch, but after being stopped and turned around, he would strike a fast gait in a few strides, and I feel confident that he never trotted faster. Should this favorable conduct be a permanent regeneration I have the utmost confidence in Anteeo proving that tips can be carried fast at a trotting gait. Everyone who sees him admits that his legs and feet could not be in better condition, and that this is owing to his foot never having been hampered with a shoe I implicitly believe.

While the history of Antevolo, brother to Anteeo, has not so direct a bearing on the question of shoeing as that of the older, there is still a lesson that is proper to repeat in connection with the other illustrations. Antevolo was foaled on May 12th, 1881, his birth-place being the celebrated breeding-farm, Palo Alto. He came all right, and the first time I saw him, May 19th, I considered that he was as good-looking and as well-formed a colt as I had ever seen in a trotting-bred one. He ran with his dam on the foothills, and was so full of life and play that he was continually galloping.

There being a good deal of gravel and hard ground, he wore his feet away, the near one being so badly broken that the toe and part of the sole were worn entirely through. The foot was so sore that he could not touch it to the ground, and he either went on three legs or walked on the front part of the ankle joint. It appeared as though it would be an act of mercy to kill him, but Frank Covey, knowing how highly he was valued, gave him every attention.

There was quite an improvement; though, when he was brought home, December 21, 1881, it was the opinion of every one who saw him that he would be entirely worthless for anything but a stock horse.

The foot had certainly a bad appearance. The horn, in place of having a natural angle at the toe from the coronet, inclined in the wrong direction, so that the toe was nearly on a straight line with the cannon bone. Though the heel was very high, it did not touch the ground, and there was an enlargement above the coronet like a ringbone, excepting that it was restricted to the front. The day after he came home I cut the horn away at the heel until it was no higher than the frog, and when clearing away the horn at the toe, which was turned under, overlapping the sole, the blood poured out in a stream, and I then discovered that for a space of at least two inches there was a gap half an inch in width between the sole and the wall. In outward appearance there was very little resemblance to a natural foot. The toe was straight across without any curvature, and the wall on each side where it came to the ground was also straight. At the coronet the heel was abnormally wide.

When the foot was pared I drew the outline of it by holding a piece of pasteboard against the sole, as he could not stand on it; and, when returned to his stall, he limped back on the other legs, holding that one up.

My intention was to make a tip that would project as much at the toe as would make the bearing the same as if the foot were natural, but after seeing the state of his foot, I came to the conclusion that it would not do to attempt putting anything on it that required nailing. I made a boot something like a "soaking boot," but I did not use it, restricting the attention to keeping the foot clean. On the 5th of January I applied the biniodide of mercury preparation to the coronet. The 14th of January I again cut away the horn at the heel, and turned him in the small lot for an hour or two, and this was continued daily until the 26th, when I repeated the blister. This treatment was continued, and it was apparent that the blister was not only reducing the enlargement but also stimulating the growth of the horn. By the 29th of March there was such a decided improvement that I concluded to commence his education, and led him by the side of X X, and the entry in the journal at that date is: "He astonished me at his readiness, the first time he was ever led by the side of a horse, and at the trotting gait he exhibited. He is a trotter surely." I led him three times, but fearing that it was too soon to take chances of the foot becoming sore again, he ran in the lot part of the time until the 15th of May, when the harness was put on him, and after becoming accustomed to it, by wearing in his stable for some days, he was driven without any vehicle. On the 24th of May he was hitched to the breaking-cart, and on June 5th he was driven to the track for the first time, and on the 8th he trotted quarters in 59 and $59\frac{1}{2}$ seconds. On the 11th of June he was fourteen hands and half an inch high; on the 14th he trotted a quarter in 54 seconds; the 26th he made the same distance in 50 seconds. There is no necessity for giving his work and performances in detail, further than to state that he also had the pinkeye, which threw him out for a time, and owing to the same causes which prevented me from driving Antceo, he was still more neglected. At the Golden Gate Fair he won the purse for yearlings, trotting in 3:07; and in the Embryo he was second to Dawn, trotting in 3:02; the time of the winner 2:59. Since then have driven him quarters in $42\frac{1}{2}$ and 43 seconds, a half mile in 1:28, and in all these instances he was barefooted. I felt that he might trot faster with tips on his front feet, and at one time thought that I would be compelled to put on three-quarter shoes behind, as he was inclined to strike the coronet and needed the protection of "scalping boots." I found a method, however, of fastening these, at times, necessary adjuncts, and I was loth to change the treatment which had proved to be so beneficial. The ailing foot is now nearly as perfect as the other, and I have the utmost confidence that in another year it will be entirely right. All that it lacks is a trifle of the roundness of its mate, and few who are not aware of the previous ailment notice the difference.

But the rainy weather compelled driving on the macadamized streets, and there being some wear at the toe, two weeks ago (January 13) I put tips on, weighing three ounces each. The next day I drove him on the track, and I thought he showed a forty gait, though the sharp edge of the tips wounded the hind pastern above where the scalping boot came, and after that he was inclined to hitch. A "speedy cut" attachment remedied that, but again the rain came, and I have not driven him since. The *Turf, Field and Farm*, commenting on the use of tips, and referring to me driving Antevolo barefooted, asked the questions which follow:

"From this (a statement of the trotting of Antevolo) it will be seen that, notwithstanding his advocacy of tips, Mr. Simpson trotted his own colt, on at least two occasions, 'barefooted' and 'without shoes, tips or weights.' Has he, too, found that tips will not answer? If not, why did he not use them on the occasions referred to?"

These questions are answered by the short history; though, if nothing had been the matter with the feet, the experiment was worth trying, and if there had been soft roads to drive upon I would have carried it further. I have not the least doubt that Antevolo would have been capable of beating 2:40 in his two-year-old form if still kept barefooted, and with a very good chance to trot a good deal faster than that. While I claim a decided superiority for tips over the full shoe, I am not prepared to say that the foot can be kept as perfect as when without anything upon it. Still, however, as tips are the nearest approach to a barefooted condition, the benefits of the latter can be rendered available as nearly as the duties of domestication will permit. Therefore the success of this barefooted colt is additional proof that "tips will answer."

Before leaving the history of these two colts I may be pardoned for diverging from the subject under consideration to call attention to the "glorious uncertainty" attending the breeding of horses. These two brothers are in many respects entirely different. In form, the elder is immensely powerful all over. Quarters, gaskins, loin, shoulders and arms are covered with masses of muscle, and bone and tendons are in proportion. His legs are short, and he stands squarely upon them at all times. He is almost a fac simile of his sire, though his head and neck are larger, and he has rather more length. He is a trifle over $15\frac{1}{2}$ hands, still growing, and when mature will probably be an inch taller. The younger has more quality, "rangier," head and neck as fine as a thoroughbred, lighter limbs and smaller feet. He is $15\frac{1}{4}$ hands now, at least two inches taller than his brother at the same age, and he has grown $4\frac{1}{2}$ inches since the 11th of June, and is likely to be 16 hands before the 1st of January next, and while he still ranks as a two-year-old. This growth also proves that the work was not a drawback.

The younger has more knee action and a longer stride. But the greatest difference is in the disposition. That of Antevolo could nôt be improved. There is not a point I would care to change; he is as "level-headed" as an old campaigner, and nothing throws him off his balance. The Berkeley railway runs within a few feet of the northern turn of the Oakland track. In the Embryo trot the train was met when at the nearest point. Dawn made a few jumps and I must acknowledge that I was in hopes that he would act badly so that I could pass him. Antevolo looked at it for a second and kept trotting his best, never leaving his feet for the whole mile. He is free and full of spirit, a chirrup sending him along while a word will restrain. He is a model road horse, never shies, and goes with the vim of an old horse. He reminds me of his grandsire, A. W. Richmond, and still more of his great grandsire, old Blackbird. The mischievousness of Anteeo undoubtedly resulted from the petting when a colt, and this trait may have led to his stubbornness as well. Still, as it was a peculiarity of Bonnie Scotland, and, as I have shown, of some of his colts, it is likely that there was an inherited tendency in that direction which might have remained latent under better management. The maternal grandam of these colts was Columbia by Bonnie Scotland and she was a grandaughter of Fashion by imported Trustee. Thus the thoroughbred predominates on that side of the house, and their action is that of the thoroughbred. As one has done fairly well with tips, the other still better barefooted, the inference is just that weight is not necessary for that kind of a gait, notwithstanding that such is the general opinion of experts.

CHAPTER XIII.

ENGLISH IDEAS OF HORSE-SHOEING — TIPS AND CHARLIER SHOES.

Having preserved a few copies of the *Field*, the great English authority in matters pertaining to breeding domestic animals, stable management, etc., containing short articles on shoeing, with the intention of incorporating them in "Tips and Toe-Weights," a portion of them is given here. These were published in January and February, 1881, and the views in places are so similar to those which we made public some seven years ago, that it was very gratifying to find them corroborated. The Charlier method is simply an improvement on the Goodenough system, which the author introduced into England nearly twenty years ago, the improvement consisting of restricting the metal to the width or rather thickness of the wall or crust of the hoof, and using a much lighter shoe. Then it is put on with a much greater degree of nicety, the horn only being cut away where the metal takes its place.

Should there be any necessity for a full shoe, this is assuredly a better plan than incumbering the foot with a load of iron, unless there is proof that weight on the heel is a requisite for fast trotting. Although we are not yet prepared to say authoritatively that weight is not needed on the posterior portion of the foot, all of the experiments so far have led to the belief that it can be dispensed with. There may be horses which will be benefited by weight on the posterior part of the foot, and if so, a properly made bar-shoe will be the kind to use. We are satisfied that the bar, or round shoe, owes whatever merit there is in it to the pressure on the frog; when this is effected more in accordance with nature, there is a greater gain.

The Charlier is superior on account of its near approach to a natural state, the whole thing being in replacing so much of the wall with a harder material. That it cramps the heel, is also apparent, though far less than the ordinary shoe. The following article is from the *Field* of February 5th, 1881:

During the last fortnight of Arctic weather the state of the London streets has been a subject of interest to all and sundry, whether they walk or drive. Various feeble efforts have been made with the object of giving foothold to the unlucky horses whose mission it is to drag heavy weights over a surface whose quality of material varies in about every other street. Here we find a shovelful of ashes; anon we splash into a "freezing mixture" (*vide Lancet*) of salt; next small curling stones (called by courtesy gravel) are sent spinning along the frozen surface by the foot which they are supposed to benefit. Now, there is no sort of reason why horses should not be able to travel over the streets in safety, whatever may be their condition. It is merely a question of rational shoeing. In the ordinary mode of "roughing" a horse, the heels of his shoe are turned up, and his foot is thereby prevented from sliding forward on ice, consequently he can stop or even back his load; but when he wishes to start, and in pursuance of his intention attempts to dig his toes into the ground, his feet fly from him in every direction, and he either falls heavily, or escapes that fate by sheer good luck. The antidote to this evil is simple enough, and why it is not universally applied is a puzzle to me. One would think that the numerous veterinary forges of London ought amongst them to be able to shoe a horse for frosty weather; but, judging from results, the contrary appears to be the case.

The whole secret of traveling over ice may be expressed in three words, viz.: "Rough the toe." A catch at the toe of each foot is perfectly effective. With three such catches, one at the toe, and one at each heel, a horse is independent of weather. Snow will ball in a foot so shod; but in towns that is seldom a very serious consideration, as when it balls it is more or less in a state of thaw, and as a consequence disappearing; while, if it be very deep, the balling does not much matter. Snow will under no circumstances ball in a foot shod with a leather sole, nor in a foot shod with a short tip, it being the heels of the shoe that perfect this undesirable manufacture. In theory the catch at the toe may be objected to, as being likely to make the horse stumble. In practice it has not that effect, as all the horses in Canada can testify. These remarks apply to harness and draught horses, whose toes are bound to come to the ground every time they start their load. Saddle horses take a sufficiently firm hold of the ground with short tips in front; and the hind shoes should have a catch rather on the outside of the toe, to obviate danger of tread and overreaches.

Omnibus horses should, in common with all others who drag heavy weights on slippery stones, let alone ice, have the toe of the hind shoe *square*, and a broad catch to it, the toe of the hoof projecting in front. The Midland Railway Company shoe their cart horses thus, and with the best results. When the foot is brought up "all standing" by a toe calk, "there it is!" When by calkins at the heel, the back sinew is too often strained. The observant will see on cab ranks many horses whose heels behind do not touch the ground at all; this contraction of the sinews is the result of the above form of sprain. My brother sportsmen must have personally experienced the difference between catching the toe, and the heel, of their shooting boot upon an unexpected obstacle. A few moments' discomfort, and, with the possessors of ill-regulated minds, a possible malediction, see them through the first casualty; while the loss of a day's, or of several days' sport, may be the consequence of the latter. Hunters in this sort of weather are on the straw ride; but if shod with short tips, they may go anywhere on the roads.

We have lately heard a great deal about working horses without shoes, but the writers on this subject have been almost to a man theorists. If anyone had given his own experience, his testimony would have been interesting, not to say valuable. "Free Lance" appears never to have practiced what he preaches in his book, "Horses and Roads," although one of his correspondents used a barefooted pony. One gentleman wrote to the *Daily Telegraph* to the effect that he drives a barefooted horse eight miles daily. Now, I know by experience that light work—say eight or nine miles daily—may be done by a barefooted horse, supposing his feet to be good hard ones; but that is mere exercise. Work is another thing. In no known part of the globe are horses able to work unshod, at any rate in front, on hard ground. In ancient times slippers were worn by horses and mules before shoes were invented. The want, though not supplied, made itself felt.

Nowadays we hear a great deal of South American horses, Indian ponies and the like. As a matter of fact these unshod horses inhabit districts where they never encounter a stone. Put them to work on rocky ground, and they are either shod or lamed. A friend of mine, lately returned from America, tells me that, though the horses were unshod on the plains, when a march over rocky ground (the Andes, for example) was contemplated, the same horses were shod with a shoe of raw oxhide. These shoes last about a week, by which time the mountains are generally crossed, and the soft plains regained. Pack mules seldom, if ever, require this protection to their feet; but even with them exceptions exist to prove the rule. And apropos of barefooted steeds, I may observe that they decidedly slip more on greasy ground with a hard substratum than horses with shoes or tips. A barefooted horse is far more pleasant to ride, to my mind, than a shod one—so long, that is to say, as he can go comfortably; but I do not know that he is more pleasant than when shod with light tips. I have not, however, given a fair trial to tips behind, although I mean doing so, as what experiments I have ventured on have been satisfactory.

To return to the insecure foothold of the bare foot. Example: In the beginning of the present season I was riding a horse out hunting with no shoes behind at all. The day was pouring wet after a spell of dry weather; the country, an ex-rural provincial. My troubles began at a bank, my horse's hind feet flying somewhere under his girths, and landing him on his tail, in luckily so trifling a ditch that, although we emerged with a most unbecoming scramble, we did emerge without dissolution of partnership. I tried to console myself with the thought that my flier had a soul above cramped banks and ditches; but, shortly afterwards, in crossing a common at best pace, where the other horses all slipped about very much, my steed suddenly came on to the floor bodily, and worse, on to my leg, his hind legs having again gone from under him. And this sort of thing, having gone on in a modified form all day, set me to thinking. As aforesaid, the ground was hard below, and very greasy atop. The horse is a flat-footed one, and this was, of course "against him," as the phrase goes.

Hunters shod with short tips go so much better through dirt than when they wear full shoes, that I am compelled to believe that their heels, being free, expand on coming to the ground, and contract when the time comes for dragging them out of holding clay, etc. I have not been troubled with an overreach since I have used tips, and fancy that the fore foot gets more quickly out of the way of the hind ones when thus shod.

Nimshivich asks (*Field*, Jan. 22): "What is the condition of frog which enables one horse to go with a five-inch tip, and another only with the whole shoe?" The answer is, that a thrushy frog, if badly diseased and in a state of inflammation, is in too tender a state to withstand stones, etc.; and I once knew a case in which the sensitive frog was all but exposed, the horny frog, or what thrush had left of it, being worn down. The animal, a four-year-old mare, very soon recovered, and had sound frogs until a railway accident destroyed her at six years old. All sound horses will go sound in short tips; but they will not thump their feet down with the action so dear to London coachmen, as if they wished to penetrate the pavement with their legs. They will use their knees less, but their shoulders more. Thus shod, I have never had an instance of a horse breaking his knees on the road, with one exception, and he slipped on a piece of ice when being led at exercise. His shoes were full Charliers, and he broke his knees through his kneecaps. Had he had tips, I think he would have stood up. The full Charlier shoe is not such a safeguard against slipping as the tip, which puts the heels as well as the frog on the ground. Now, I must disagree with "Nimshivich's" friend as to our forefathers being experts in shoeing. The very old English shoe, worn by the "destriers" (which we should call cart-horses) that carried the knights of old, was clumsy in the extreme. Its great width, however, allowed some pressure to the sole, which was so far good, as experience shows that an unpared sole can stand pressure. I can only account for the horses of our grandfathers standing work at all, shod as they were, firstly, by their being nearer to their Arab ancestors, and consequently more sound in themselves than our horses; and secondly, by the custom of turning them out to grass on every available occasion. They lost condition, went broken-winded, etc.; but their feet had a chance, or rather nature had a chance of restoring their feet; and, all the world over, there is nothing like dew and night air for anything like fever in the feet. Harness horses had no weight on their backs; those belonging to rich men were not unreasonably worked; the mail coach horses did their stage in the allotted time, and the number of legs which they used in progression was optional with them. In many ways our horses have certainly deteriorated, and in nothing more than in unsoundness of wind.

The first cart-horse I ever heard roaring caused me to turn round and make a mental note of the circumstance. Now I often hear them, and, more remarkable still, last summer a pony of the polo description cantered by me, roaring like Prince Charlie himself.

The greatest errors in the matter of shoeing history are to be found in veterinary works of a comparatively late date—say sixty years ago. What countless horses have been ruined by the theory about the descent of the sole! Our great-grandfathers and grandfathers had plenty of horses lamed by their shoeing, but they ascribed the lameness to every cause but the right one. Chronic laminitis they called "chest founder," and many a horse lame in his feet was tortured by having his legs fired and blistered. Nor is all that practice quite changed in the present day, though much improved. To return, however, to the descent of the sole. Youatt, in "The Horse," page 418, says that, unless the sole be pared, "that portion of horn which in the unshod foot would be worn away by contact with the ground, is suffered to accumulate until the elasticity of the sole is destroyed, and it can no longer descend, and foundation is laid for corns, contraction and navicular disease!" Fancy a man calling himself a director of public opinion about horses, and not knowing that the sole never attains more than a certain thickness, whether it be on the ground or not! When thick enough it scales away, leaving a new sole ready for any emergency below it. The sole, being pared, required protection, and could not even be touched by the shoe. Hence the necessity for the foot surface being beveled off, and hence contracting corns and navicular disease.

I do not say (because I don't know it) that a horse shod with tips, and having his frog as Nature made it, would never have navicular disease, but it stands to reason that he has the best chance possible of escaping it. Firstly, he minds where he is putting his feet, and does not bang them recklessly about; secondly, the thick horny frog shields the navicular bone and perforans tendon.

Otterbourne asks how Charlier shoes, in front only, answer? I reply, very well; but I prefer the nearest approach to nature all round, excepting for horses drawing heavy loads, and they have been discussed above. With the old-fashioned way of shoeing, only feet of medium quality, neither too strong nor too weak, had any chance of standing sound. Without shoes, sooner or later, all horses really worked on the hard will be lame. With tips, all horses can do all reasonable work, excepting those suffering from navicular disease. Tips will not make lame horses sound at once; but they give many diseases of the foot the best possible chance of righting themselves, especially if the owner of the horse possesses the quality of patience. Last season I owned a mare who had such thin feet that she could do nothing at all barefooted on the road. With short tips, however, she had, if anything, rather too much action for a hunter. At present I have a horse whose heels are quite on the ground; yet he goes better like this than he did with full shoes. Tips are no new notion. The "lunette" shoe of Lafosse was a tip. I do not think tips need be very narrow, unless they are let in *a la* Charlier. The heels, however, should always be beaten down quite thin. Clips are most useful in keeping light shoes or tips in their place; but they should on no account have a place cut out for them in the horn. Let them just be hammered on the outside of the crust. Also let the clinches be cut off and knocked down, but not rasped, as the crust is rasped with them, and a chance is given to the nails to break out. Let the sole and frog be on no account touched with a knife. Shorten the toe as much as possible on the occasion of each shoeing, with a rasp. And above all, "gentle reader," if you don't happen to be master in your own stable, never let the words "Charlier," "tip," or "nature" escape your lips. If, however, you are not a slave to your groom, my experiences may be of use to you, and in any case are very much at your service.

P. S.—The above remarks having been delayed in transmission, I take advantage of the opportunity to add some more last words, and to make some remarks, which I hope will not be considered impertinent, on the letters written by various gentlemen *in re* Horse Shoes which appeared in last week's *Field*. First, Waverly, who was persuaded out of the Charlier system by his farrier, says that I take for granted, for reasons which I do not attempt to explain, that grooms and smiths are opposed to Charlier shoeing. I speak from experience, as all the grooms, excepting two, with whom I have ever discussed the subject, are opposed to it, for "reasons which they do not attempt to explain." As to smiths, no less than four of them have refused to shoe my horses at all on this principle; they have all been owners of superior forges. I never had my horses better shod than by a village blacksmith who did what he was told; but even he charged me 2s. per set more for tips even than for the old heavy shoe! I may say as regards Waverly, *De te fabula narratur*, as Mr. Fardous speaks his own sentiments. Had Waverly persevered with the shoeing, the discolored horn would have grown and been worn out; his horses, not being lamed by the original bruise, would not have been lamed later on, any more than mine were.

I agree with M. B. that Charlier shoes, not approaching the corn place, cannot cause corns, any more than a gag snaffle could give a sore back. I should like, however, to know whereabouts the soles of his horses were worn unduly thin, as the frog of a flat foot, being always prominent, shields the part "aft" of the point thereof, and the shoe should protect the forward part, it being as impossible as undesirable to have the shoe in a very flat foot flush with the sole. Is M. B. quite sure that the soles were in a natural condition? He says he never has the frog touched with a knife; but, if not absolutely watched, smiths will, under the pretense of removing " ragged dead horn," thin the sole. I am curious on this subject, as last season I possessed a hunter with absolutely pummiced soles, which improved immensely in the three months that I owned him, one foot regaining the normal appearance, although in both the coffin bones had descended when I bought him, as a stop-gap and as an experiment. I much regretted that I parted with him, but a severe accident was accountable for that. Capt. Gillon is of my way of thinking. Would he tell us whether the material for shoes which he mentions is procurable in England? He, like myself, has given the system a fair trial.

Suaviter in Modo also agrees with me, for his harness shoe is the modified Charlier, with which I have shod many thin-soled horses until the sole grew thick. Apropos of his remarks on curby hocks, I may say that several years ago I bought a remarkably clever young Irish mare, fired for curbs, and with curbs which appeared callous. At the end of a season of Charlier hind shoes, the enlargements were absorbed; and I sold her to a dealer at the end of the next. Finally, Mr. G. Johnson is specially happy in his remark, that in most forges " the quantity of iron employed far exceeds that of common sense." To turn from this well-worn subject to a fresh one, will H. H. kindly tell me whether his hunters can eat Goode's luncheon cake out hunting with a double bridle in their mouths? I find that I must make one more remark on these weary shoes. Suaviter in Modo misunderstands me as regards three nails only to a shoe. I use three nails in the *tips*, of which I have given a sketch (there is hardly room for more), and it was to these infinitesimal tips only that I referred.

HORSE-SHOEING—THE CHARLIER V. THE OLD SYSTEM.

Sir,—The views expressed by W. J. R. in your recent issues much according with experiences which I have derived from an eight years' practical use of the Charlier system, carried out under my own personal supervision, it may not be uninteresting to consider the question by the application to it of governing principles; for after all there must be, and there is, a reason for everything. For this purpose let us see, in the first place, of what does the horse's foot consist; then the reasons for applying shoes to it at all; then how far the old system and that of the Charlier effect the objects desired; and, finally, the weight of advantage and of disadvantage attending upon each of those systems.

The foot of the horse, then, may roughly be said to be constructed of a series of sensitive and insensitive stratifications, the one alternating with the other; the semi-circular outside wall or crust, the outside sole, and the outside frog, being naturally of the insensitive kind, whilst the inner semicircular crust or wall ("lamina") the inner or underlying sole, and the inner or underlying frog, are of the sensitive—indeed, very *acutely* sensitive kind. This foot, as it is illustrated with the unshod and still unimpaired colt, is strong and elastic, but solid and without concavity, the well-used and developed elastic frog filling up all the centre of the foot, and by its wedge-like operations preventing the possibility of any contraction at heel, whilst it also fully performs its valuable functions as "buffer," intended by nature for the relief of the joints of the limb from jar, and the foot itself from concussion, as the result of striking the hard ground with rapid action.

This elastic insensitive outside frog is also given to the horse that he may have a foothold upon hard and smooth surfaces, as I have seen to demonstration in the case of unshod horses carrying heavy burdens in safety over rocky tracks in the Himalaya mountains, where a shod horse could not have even put in an appearance and carried himself; and as I have also had experience upon the flat surface of Cheapside, when neither wet nor dry. It is also quite clear that the unshod horse must be vastly less liable to sprain his back sinews, the developed and projecting insensitively elastic frog constituting a wedge of support to a vertical pressure upon the pasterns, and thus minimising the consequent leverage upon the back sinews. There can be no doubt that this insensitive outside frog is highly and sensitively accumulative, or the reverse; cleanliness and exposure to healthful pressure, and of that the more the better, producing growth, resulting in its full development; whilst its withdrawal from this healthful pressure, aided by dirt, will as quickly result in its shrivelled proportions; and even, indeed, in its absolute disappearance—when, of course, contracted heels are also an inevitable consequence. These evils, however, as we have seen, cannot fall to the lot of the typical colt, when at large on his pasturage; although they may, and do, happen to him also when kept for any length of time in a dirty straw-yard, and thus deprived of his natural contact with his own mother earth.

What, then, are our motives, it may be asked, in applying shoes at all where the typical colt is thus seen to go as daintily and gracefully unshod as do the peasant girls of the south and west of Ireland, and also in many other (for the present, at least) happier countries familiar to many of us? The answer is quickly given—that the colt's earlier career does not require that he should perform rapid pilgrimages on roads strewed with sharply angular and artificially broken stones, which would otherwise risk the occurrence of either of two kinds of injury—the one being the breaking away, splitting or tearing of the outside, insensitive, protecting wall of the foot, and exposure of the inside sensitive wall or "laminæ," as might be exemplified by the breaking of one of our own finger-nails to the quick; the other injury to which he would be liable being caused by the descent of his foot with force upon one of such sharp stones, and the consequent bruising of the underlying sensitive sole, the outside insensitively protecting sole having proved itself an insufficient shield for its protection against so unnatural an assailant.

In answer to the question of how the colt is to be protected from either or both of these misfortunes—both, as we have seen, proceeding from dissimilar causes—I reply that the remedy will depend upon the position in life it is intended that he shall occupy. If to race, hunt, or, in a word, to "carry a saddle," then I unhesitatingly say, by shoeing him on the Charlier system—a narrow rim or moulding of steel protecting the insensitive outside crust and maintaining its semi-circular formation intact as well as a clumsy appendage of iron, and without its manifest disadvantages; and, as he will be free to choose his own ground when traveling by road, he requires no protection from the second kind of injury already referred to. If, on the other hand, his destiny should be fast harness work, to be performed on roads still unvisited by a steam roller, and where (especially in double harness) he is powerless to choose his own ground, then I have, by unpleasantly gained experience, learnt that the Charlier system does not afford the necessary protection, and that it is desirable to use a shoe of the old type, with a bearing, however, restricted to the outside insensitive crust only, a flatly projecting flange extending to the inside of the semi-circle, the unimpaired and developed frog having still a bearing in the centre almost flush with the flange or flat web; a knife being in no case applied either to frog or sole, and the desired diminution of length of toe being effected by a rasp, as we should file the ends of our own finger-nails.

The advantage of the Charlier system to that class of horses first referred to, and also to driving horses where steam rollers are in vogue (and at times of year when country roads are not in a state of chronic reparation), is that they are light, ensure frog pressure with its consequent development, and therefore wide and open heels, freedom from windgalls, diminution of risks of fetlock cutting, brushing, speedy-cutting, and tendon and joint straining; the foot, with the addition of a steel rim or moulding to preserve its integrity, being in all other respects in the same condition as that of our happy colt—solid instead of hollow, and therefore free from the risks of picking up stones, of shoe-pulling through suction in deep rides through clay coverts, when the familiar sounds of "cork-drawing" are in the air.

"Impecunious" observes that farriers are opposed to the Charlier system—the reason for which, however, are by no means hard to find, for it requires more delicacy of workmanship, more care and time—in fact, the services of a more skilled artisan; the *modus operandi* being first to rasp away the extra length of toe which had grown during the interval between the last shoeing or removal, as may be (and which in the case of the colt would have been constantly and regularly kept back by friction of the toe against the ground), then to lower the groove in the insensible outside crust to its proper level, and then—that which is the real difficulty—to adjust the steel shoe with perfect accuracy to the form of the foot and groove, before a nail can be driven; whilst in the old system it was only necessary that he should turn a ponderous mass after some standard pattern of his own conception, which a few slashes of his knife sufficed to make the obsequious but suffering hoof conformable to. It is also to be borne in mind that the Charlier shoe, being composed of steel, involves much more wear and tear of the tools used in its construction, especially of the rasp used in bevelling off its sharp inside edge. Assuming, however, that this shoe is put on by a man acquainted with the construction of the horse's foot, with the hammer concentrated in width of surface bearing to the insensible semicircular crust only, and that the heels are not curved inwards and drawn too long, it is physically impossible that such a shoe could cause corns. One of your correspondents suggests that in certain instances the Charlier shoe should be fastened on by three nails only, in order to avoid risks of lameness from using more; but I confess that I am at a loss to see any reason for it, as the fewer the nails used the greater must be the strain upon each; and beside there is practically a necessity for nailing the shoe at pretty short intervals, because if the horse do road work and have not unusually perfect shoulders, the toe of the shoe quickly wears thin, allowing the heels to expand beyond the dimensions of the foot to which they had been originally adjusted, where such parts of the shoe are not riveted in their original positions by the nails.

When upon the subject of the old system—which under, the circumstances referred to, would appear to be the better system for carriage horses driven over stony roads—I have made no allusion to the barbarous and cruel application of "heel cocks," as it is self-evident that if the principles of the Charlier system possess no more virtue than that which is involved in the mere question of superiority, which has led to expressions of opinions by your correspondents, such appendages must needs be wrong, and that beyond any question. Much has been well said and written against the cruelty of bearing reins, and with happy results; but if any of your readers should doubt the cruelty practiced upon a London carriage horse, let him turn into any one of the fashionable West-end forges, ask for an ordinary hind shoe, such as is kept in stock for carriage horses, lay it on the ground, and see the angle at which it necessitates its unlucky bearer to stand and move upon it, without reprieve to its distorted tendons except when lying down—the altitude of the outside "heel cock" being about double that of the inside, and the weight of the animal distributed equally between the unequal points of the one and the other, with the point of the heel for a tripod. Of frog there is none remaining, and therefore it is out of the question, of course. Some say: Very true in general, but my horse has "curby hocks," and therefore requires these "heel cocks" artificially to support them. I say, certainly not. On the contrary, knock off the heel cocks, lower the shoe at the toe, as much as nature will permit, allow the frog to receive pressure from the ground and gradually to come on the scenes again; the tendons will meantime readjust themselves and assume their natural proportions, the animal becomes at ease, and a sound and actively elastic frog will amply supply the rest.

SUAVITER IN MODO.

HOOFS AND SHOES.

SIR,—That science has made great strides in many things no one will deny; but that it has made equally great progress in the art of shoeing that very patient slave of man, the horse, no one, I venture to say, will admit. The manner in which the majority of horses are shod shows too plainly that the quantity of iron employed far exceeds that of common sense. The farrier first cuts away the sole until it will "give" under the pressure of his thumb, and then nails on a broad mass of iron to protect it. The frog, and even the bars themselves, do not escape his barbarous knife: for he no sooner finds that they have grown a little since the last shoeing than to work he goes, and again undoes that which nature herself has done. But let us not be too hard on the much-abused shoeing smith, for he often has to work according to orders from the man in charge of his four-footed friend, who will not be satisfied until the feet look like so many cockle shells; pared and rasped to the last degree is his plan. Of course they look "clean and nice," and that is about the only explanation he can give you for having it done. No wonder, sir, we see so many contracted feet. If our aim is perfection, we must work *with* nature, and not against her; and I would ask any man with a grain of common sense if the above method is in harmony with the rules laid down by nature. If it is not, then I say the sooner we discontinue such an idiotic system the better for all concerned. Why, let me ask, do we shoe horses' feet at all? For the simple reason that the wear and tear of the crust is greater than the growth. It is then with the crust, and the crust only, that we have to deal in order to make a horse perform his allotted task without injury to the foot; and if we can do that, and at the same time preserve the same even bearing of the foot which it possessed in its natural state, I think we shall then have arrived as near perfection as any reasonable man can expect. Well, I say we can do it if we put into practice Charlier's system; but it must be carried out to the very letter, otherwise we shall be subject to disappointment, and apt, like many others, to condemn that which we do not understand. I have before me a hoof— or rather an imitation of one—given to me by M. Charlier some twelve years ago, showing very distinctly how a foot should be prepared and shod on his principle, and I shall be pleased to show it to any one sufficiently interested to inspect it. In conclusion, let me advise those who are prejudiced against his system, and those who have never tried it, first to get well acquainted with it by having it properly explained, next see that his principle is carried out in every detail every time of shoeing, and then give it a thorough trial, and I am convinced that they will be more than satisfied, and regret that they did not try it sooner. G. JOHNSON.

34, Woods-mews, Jan. 26.

HORSE SHOEING.

SIR,—Will you allow me the opportunity of correcting "Mustang's" misapprehension in supposing that I had "advocated the use of many nails," as what I had intended was to state the necessity for more than three nails in fastening on a full Charlier shoe, in contradistinction to the "infinitessimal tips" only, referred to by "Impecuniosus" in your last issue. This practical necessity, as I had previously, but no doubt inadequately, endeavored to explain, arises from the fact that the closely hammered steel Charlier shoe, with road work, quickly wears thin at the toe, when its tendency is to spread out at the heels; and, unless where and so far as it is actually held in place by the nails, to extend towards the heel beyond the dimensions of the hoof—with the consequent danger of cutting the fetlock of the other leg in action, or of pulling off the shoe, by offering it as a projecting flange to catch and hold to the ground when the horse takes his foot out of dirt.

I also tried to convey my opinion that there is no disadvantage in thus using any reasonable number of nails for the purpose; and I confess that, notwithstanding "Mustang's" explanations that the "crust is composed of fibers running parallel from the coronet downwards," such is my opinion still. For in the same way our finger nails are composed of fibers running parallel from the quick downwards; and yet the quick is unaffected by the number of small holes which may be carefully drilled into the projecting ends, so long as they do not cause splitting or injury upwards. The Charlier shoe, being itself excessively narrow, necessitates that the nails used in fastening it on should be driven correspondingly near to the edge of the insensible semicircular wall or crust—so near, indeed, that the growth of a healthy hoof enables the margin, which had served as the receptacle of the row of nails at one shoeing, to be altogether done away with by the rasp at the next monthly shoeing or removal, as the case may be. But as I have already intimated, the whole system of the Charlier shoeing involves more delicacy of manipulation, and a finer and more highly wrought kind of nails into the bargain.

SUAVITER IN MODO.

SIR,—Were it not for the knowledge that your liberality in opening the columns of *The Field* to such an extended discussion on horse shoeing has been attended with an immediate and practical effect, I should not have presumed to trespass on your space with another letter; but as I am aware that five of my hunting acquaintances are having their horses shod *a la* Charlier, and that three country forges, where the process was hitherto unknown, have now the necessary appliances, and the owners are prepared to devote their best abilities to the operation, I am emboldened to send you some additional remarks, in hopes of furthering the good cause.

It seems to me that the one thing needful to effect a complete change for the better in horse shoeing all over England is that veterinary surgeons should take and express greater interest in the subject: that they should, in fact, discriminate openly and strongly between good shoeing and bad, between shoeing which will probably enable a horse to remain sound in his feet to a good old age, and that which may fairly be expected to render his life miserable and his death premature. Few men can doubt that if veterinarians were generally convinced of the truth of the Charlier system, and used their influence to promote its adoption, a few years would see it introduced throughout the length and breadth of the land. That they do not do so is not necessarily a reproach to them; there may be inconveniences and dangers attendant on it, of which outsiders like myself are ignorant, but which are fully apparent to their more instructed minds. What surprises me is that they should not give us a hint on the subject.

For many weeks you have set aside a large amount of your space in order that horse-owners might ventilate this question, and try to arrive at some conclusion likely to benefit the most generous and courageous of all animals; but not one single useful contribution has emanated from a member of the veterinary profession. If it were a question of bad water or bad smells, doctors would be found in plenty to tell us what to do; but the sphinx was garrulity itself compared to our veterinarians. Surely they must thoroughly well know whether the Charlier system is good or the reverse. From the fact that certain eminent members of the profession have written approvingly of it, and from the eagerness with which one or two of your correspondents, themselves veterinarians, have hailed its inventor as a *confrère*, one would suppose they regarded it as entirely good; yet, can any horse-owner remember one of the rank and file of the profession advising him to try it? I certainly cannot; and it is this prejudice, indifference—call it what you will—on the part of so many members, that makes me consider them responsible for much of the cruelty and injury daily inflicted on horses in most provincial forges. They have no right to seek to shelter themselves behind the honored names of professional brethren whilst tacitly conniving at, or actively perpetrating, mutilations on horses' feet, that would fill with horror the men whose names they invoke, and which their own knowledge and reason utterly condemn.

That veterinians may be stirred up to a more active interference in this important and humane question is my only object in writing this letter; and no greater mistake can be made than the supposition that there is any desire on my part to deny credit to the eminent men belonging the profession, to whose investigations, not only on this but on many kindred subjects, the world owes so much.

W. J. R.

Sir,—W. J. R. might have saved himself the trouble of inditing last week his little treatise on sarcasm, with quotations from Artemus Ward and Sydney Smith thrown in. Spite of being handicapped by my nationality, his sarcasm did find its way to my mind. This, however, in no way necessitated that he should malign himself, if indeed he knew better, by appearing to judge of a horse's foot-soundness whilst it was going across country. With regard to the animals involved in his pettish sarcasm, I may state that I never rode better hack-hunters than I did during the five years I was at Oxford, and do not recollect more than two occasions on which I saw a lame horse out.

"Impecuniosus," M. B., and Mr. Tozer agree that "the Charlier shoe, not approaching the corn place, cannot cause corns." Neither can any other shoe, though the manner of shoeing may be answerable for them, inasmuch as it may cause the heels to contract and lose their power of expansion. The concussion is greatly increased, and the result is the rupture of bloodvessels in the sensitive sole—that is, so-called corns. In cases of weak feet, corns are more often produced by stepping on stones. The adoption of the Charlier would certainly increase danger from this source for the first few weeks, but would eventually result in a healthier foot, I have no doubt.

Does "Frog" speak from personal experience when he says that horses are worked barefoot abroad? In the case of the most horse-loving of Continental nations I can assert the contrary. I have visited every province of Austria, and found the horses universally shod with a full shoe. Hungarian and Polish horses, if any, should be able to go barefoot, owing to the practice that largely prevails of allowing the foals to run alongside while the dam is in harness. Whilst living in the house of a horse-breeder in Transylvania I have seen two mares harnessed to the conveyance, and, when fairly under way, found we were accompanied by six young colts and fillies, from rising one to rising three years old, each of the matrons claiming three of these, with the prospect of soon increasing the number. With a view to testing in one case what appears to be still a theory, I have had the shoes taken off one of my nags, a thoroughbred mare that I use solely for hacking, and shall try the effect of riding her without shoe or tip. BORDERER.

HORSE SHOEING.

SIR,—With your permission, I will endeavor to give "Mustang" the information he desires respecting the imitation hoof shod *a la* Charlier, by forwarding you a photo of the same which I have just had taken, feeling sure that "Mustang" and the gentleman from whom I have received letters asking for particulars will be able to learn more from it (if you will kindly give a sketch of it) than from any description which I could give with pen and ink. I may, however, remark that it is a true model of a foot shod at M. Charlier's forge, on his principle for six months, and showing how well developed the frog, bars, and sole will become, if only left alone. With a strong, thick crust the shoes may be shortened a good inch, and especially when a horse has to travel over slippery roads. Thus modified, I prefer them to tips, although until the foot has attained something like its natural state, I am a strong advocate for the full Charlier shoe, which, being of uniform thickness, and with the groove made deeper at the toe than at the heel, preserves the latter from undue wear, and consequently encourages its growth. It will be seen that a narrow space is left between the heel of the shoe and the rim of uncut crust to allow of expansion of the foot (a most essential point), and it is in neglecting to provide for such expansion that has caused many to form an unfavorable opinion on what they took to be *la système* Charlier. Now, with regard to tips, by all that is just let them be sunk into the crust, otherwise you destroy that even bearing which it is so desirable to preserve. No matter how thin the heels of the tips are made, there still remains the fact that the heel is not level with the toe. How can it be? The whole circle of the foot, whether shod with tips or full Charliers, should be uniformly level, that both heel and toe may take their equal share of pressure at one and the same time. It is nature's own law; let us not try to improve it, or we shall be losers in the end, depend upon it.

When I see a set of shoes weighing nearly twenty pounds, and with sufficient iron in the calkings to make a decent set of shoes, I feel ashamed when I say that the wearers are to be found in this, the greatest "horsey" country in the world. Depend upon it, if horses could be given the power of speech, they would call us by names which were not given us at our baptism. Then away with calkings and broad, heavy shoes, and let us give the poor brutes a chance of using to advantage that which nature has given them.

"Snaffle," in last week's *Field* asks whether there is anything to prevent the crust being cut to receive the shoe the first time of shoeing. I will give him M. Charlier's plan. When he had a strong, sound foot to shoe he would sink the shoe level with the inner circle of crust, the second or third time of shoeing, but with a weak one he would work with the growth of the foot. As it improved so would he lower the shoe into the crust, and thus by degrees bring the sole, bars, and frog into contact with the ground. Such was the plan of the great Charlier himself, with whom I have spent many happy hours, for he was never tired of endeavoring to show the superiority of his system over every other, and it is only by following in his footsteps that we can expect to arrive at anything like perfection in the art of shoeing. It would be interesting to know if any of "M. B." horses "forge;" if so, I suspect that is the cause of the thin soles (if in front). I have known cases where forgers have been ridden or driven fast to wear the soles of the feet quite thin by the forward action of the hind feet. G. JOHNSON.

34 Woods-mews, Park-lane, W., Feb. 9.

With these quotations, which embody the gist of the arguments in the *London Field* of two years ago, this portion of the volume will be brought to a close for the present. Still experimenting, I hope to add to the illustrations when another edition is demanded.

In the meantime, the subject will be duly considered in the BREEDER AND SPORTSMAN, and as the experiments progress the results will appear.

I am happy to state that others will give the tips a fair test, adding their experience to mine. Their communications will be published in the paper, and extracts from the British journals bearing on the question will also be given. Being so fully impressed with the benefits that will follow a more natural treatment of the foot of the horse, and the improvement that will follow the more rational practice; ignoring the "protection" that invariably results in injury, I feel that too much attention cannot be given.

[*APPENDIX.*]

AN ESSAY ON
TOE AND SIDE-WEIGHTS

CHAPTER I.—ACTION OF THE RACE-HORSE.

Frequently when horses are the topic the action is commented upon, and nearly every one who makes any pretentions to equine lore is prone to commend or condemn that of the animals which are the subject of discussion. The neophyte is puzzled to understand a good deal of the phraseology which horsemen use to convey their meaning when speaking of peculiarities in the animals, and a recourse to the dictionary fails to aid him in knowing what the jargon means.

Adjectives expressing every degree of quality precede terms which are inexplicable, and he ponders over the matter in amazement. Bold, prompt, true, slovenly, slow, rapid, scrambling, dwelling, round, smooth, and a score or two more of opposite or synonymous meaning are used to express the various methods of progression which horses display. There are some in common use which are still more intricate, and he hears of daisy-cutters among the race-horses, and open-gaited trotters, without having the least conception of what is meant by the obtuse designations. As he improves in knowledge and becomes familiar with the technical language employed, the uncouth phrases have a meaning, oftentimes very expressive and appropriate, giving a lucid, if terse, explanation more effective than long descriptions.

In a treatise intended to account for the effects of toe and side-weights, action is the first thing to consider, and if we cannot find a key to the problem in this study, it will be useless to look for it in any other direction.

Long before the era of these latter-day appendages to the feet of the trotter, it was well known that the action could be modified by artificial appliances; and so long ago that it has become dignified as a proverb, there was a saying that "an ounce on the heel is equivalent to a pound on the back," the implication being that weight on the feet influenced the action of the race-horse prejudicially, so that he would tire quicker than if he had to carry sixteen times as much. Something of the same idea governed when it is said that one horse could beat another "shoes to plates," as this was about the strongest term to convey decided superiority which could be used.

It will not be out of place, then, to give some consideration to the sources of action, so far as can be shown by the configuration; and yet the form will not decide, for in exceptional cases the horse of the truest proportion may be faulty in his movements, when the one of inferior shape is the superior. Were it not so, there would be no use in the endeavor to remedy defects by education, for if the form absolutely governed, that would end it, and the only recourse would be to perfect the form by breeding after the desired type. The modification can be accomplished, and the slouch is transformed into the graceful dancer, and the members of the "awkward squad" become models of precision.

In training a man, his mentor can explain to him wherein his "style" is faulty, and the necessity for acquiring a better method of using his limbs.

All the trainers in the world could not change the shamble of Weston into the perfect gait of O'Leary, but if the education had been commenced in time, there would have been an approximation to the desired end. Reasoning and example are the dependence, when man is the pupil, with practice to perfect; in the horse, mechanical devices take the place of precepts, and the combination of these, and the animal's natural intelligence, are all that the trainer has to aid him. But it is perhaps premature to consider the resemblance between the *style* of the man and the *action* of the horse now, as they will come in more appropriately hereafter. Neither is it necessary to give elaborate attention to the anatomy of the horse, and a brief review of the general structure and physiology of the animal will be sufficient.

The bones are the mechanical part; the muscular system a portion of the motive power. It is generally conceded that the brain and the nerves which spring from the brain furnish the balance of the force, and it will be all that the present purpose requires to accept this as the correct theory. It may provoke a smile when I claim that the action of the brain is affected by mechanical contrivances, and yet I shall fail in my own estimation if I do not make it apparent. The skeleton of the horse is tolerably familiar to those who have given any attention to the formation of the animal, and if it has not been studied when divested of all the tissues, a very good idea is obtained from cuts and engravings.

This frame has some similarity to that of man; in other respects the difference is strongly marked. Though all parts have more or less to do with progression, the limbs and lumbar vertebræ are the most actively employed. A glance at the skeleton will show that the scapula, shoulder-blade, and the humerus, or upper arm, form quite an angle. At the junction of the humerus and radius, the elbow projects upward to a greater heighth than would be thought from looking at the living animal. Projecting from the back part of the knee is a bone, termed by some of the writers the trapezium, and this is more prominent in the skeleton than would be supposed from its appearance in life. At the ankle the external sessamoid has a backward prominence, and from the elbow to the foot each joint shows that the greatest force is required from behind, the attachments for the tendons and ligaments being so much larger than in front. When these joints, as in the upper parts, are so thickly covered with muscles, and rigidly bound with ligaments, it is evident that there cannot be much side motion, the bending being more like the straight working of an ordinary hinge.

When the foot is raised it is thrown outward a little, unless there be malformation, and when elevated until it touches the elbow, the frog will be outside of the ulna. This is caused by the slight curvature outward of the bones. The purpose of this is evident, as it reduces the jar by bringing the concussion oblique on the joints. Were the supports a straight column, it would not be long until the padding between the bones would become inflamed, and in a short time the bones would be diseased. The bones of the posterior part of the frame are similar in respect to angles and attachments for the muscles, although we find a greater degree of obliquity. The angle between the os innominatum—the bone which forms the slope of the hip— and the femur corresponds with that of the scapula and humerus; but from the stifle to the hock there is a reverse angle, and from the hock to the foot the inclination is again to the front. Thus there are three pronounced angles behind to one in front, and the old writers explained the necessity of this configuration to the bones being connected with articulations, whereas the front was guarded by the scapula being only attached to the frame with bands of muscular fibre.

The instantaneous photographs have corrected many of the old-time errors, and though the deduction was correctly drawn in that particular, a great deal of the reasoning has been upset. Thus it was thought that the animal received the first shock, after the air-flight, on the front leg, and the yielding of the muscular attachments of the scapula to the body was to correct this. In the leap proper, like going over an obstacle, this is correct; but in the flying gallop the first contact is with the hind foot, and in the square trot the hind and fore feet strike the earth so nearly together that the photograph fails to note anything but the slightest difference. If difference there is, it is in the first touch of the hind foot. But before entering on the interesting lessons of the instantaneous photographs and the fund of knowledge they present, it will be as well to consider more fully the frame, its covering of muscles and the tendinous and ligamentary attachments. There are immense muscles enveloping the quarters. extending over the loin and following the back-bone. They form protuberances, called by horsemen the gaskin or lower thigh, and in some horses nearly fill the space between the ham-string and tibia.

In places these are attached by tendons to levers, connected on the fore leg to the elbow, trapezium, external sessamoid and pedal bones, and in the hind leg to the patella, calcis—point of the hock—sessamoid and pedal bones.

These attachments have a great deal to do with locomotion. Anatomists group the lower muscles as flexors and extensors—the office of the former being to flex the limbs by pulling them up, the latter to extend them and thrust the foot forward; and the tendons which convey the force to the extremities are also called flexors and extensors. It seems to me that these terms have caused a misapplication of the uses of these motors by those who have only a limited knowledge of the functions of the muscles, and have associated the most powerful with the duty of merely flexing the limb. This is an erroneous conclusion, for the flexors of the fore legs give the last and immense propulsive effort to hurl the body through the air in the fast gallop; and though the fast trot depends more on those of the posterior limbs, all are called into the service of progression. The flexors are elongated when the foot is thrust forward by the contraction of the extensors, and the drawing up of the fibres throws the body along.

To properly understand the effect of toe-weights on the trotter, it will be necessary to give attention to the action of the race-horse, and proving the truth of that which seems at first paradoxical, that weight is a drawback to one, an advantage to the other. And now, without the aid of the instantaneous photograph I should be completely at fault, without cue or scent to guide me on the trail. To be fully understood by readers, it will be necessary to get the cartoons published by Muybridge, for words are inadequate to give a proper understanding of the subject. The illustrations of the fast gallop are on one card, there being eleven pictures of Sally Gardner, the eleven covering one stride. The cameras which caught and recorded the shadow of the flying animal were placed twenty-seven inches apart, the slides being opened and closed by an electrical apparatus, so that there was only an exposure of less than the two-thousandth part of a second, the whole stride of 265 inches being made in forty-two hundredths of a second, and each camera passed in thirty-eight thousandths of a second.

It is difficult to conceive such rapidity in taking and recording the representations, though any one who will take the trouble can convince the most skeptical, by a mathematical demonstration, of its truthfulness. The breaking of a thread opened and closed the slide, so that the question is merely a simple rule-of-three proposition of: If an animal is covering 633 inches in a second, how long will it take to pass over a space represented by a thread the twentieth of an inch in thickness? The answer gives 12,660 in a second of time, so that Mr. Muybridge's claim of a two-thousandth part is far inside of the truth.

As we progress in the consideration of the action, as disclosed by the instantaneous photograph, it will be evident that this celerity was absolutely essential, or the whole thing would be a confused blur.

The background of the cartoon shows a screen divided into a number of vertical spaces twenty-seven inches apart, numbered from one to seventeen, and four horizontal lines, the lower being the ground surface, the others four inches apart. There is also another horizontal line which marks about the height of the mare, nearly touching her croup and withers when she is at rest. If the picture at rest was taken at exactly the same distance from the camera as those when she was in motion, it will prove another new feature in the stride of a race-horse, viz: the height which the body is thrown when in the air.

Number one of the series discloses the left fore foot on the ground in advance of the line between the spaces five and six; the pastern is bent so that the ankle nearly touches the ground. The vertical line is midway of the ankle joint, bisecting the body about four inches back of the elbow, and, as nearly as can be told, the middle of the saddle. The body of the rider is thrown forward so that the upper part of his back is in advance of the line, his head, even the back portion of it, being six inches in front. The left fore leg is bent at the knee, the front part of the knee being twelve inches from the ground, the heel of the upturned foot eighteen inches from the ground. The hind legs are bent at the hock, the toe of the right-hind foot touching the line which marks an elevation of twelve inches, about twenty inches in advance of the left hind foot, which is still more elevated. The metatarsal is exactly vertical, as shown by the line between the spaces three and four. The horizontal line answering to the height of the mare, shows above the withers, consequent upon the bending of the pastern, and is also above the croup, the elevation of the hind feet being occasioned by the bend at the hocks. Thus the whole weight is sustained by the left fore leg, and nearly in the position where the last propulsive effort is made to send the body along. As number eleven of the series presents a better view of the

"LAST EFFORT,"

I will defer remarks on that until it comes up for consideration. In number one the nose was exactly touching the line between spaces seven and eight; in number two the nose is advanced a third of the way across space nine, showing that thirty-six inches had been covered, owing to the thread being wrongly placed or carried along a short distance before the tension was sufficient to break it—probably the latter being the cause. This picture is a complete refutation of the erroneous conception heretofore prevailing among artists who persist in delineating this part of the stride of a race-horse in the most absurd manner, fore and hind feet being thrust out in impossible positions. "Stonehenge" and others discovered the absurdity, but fell into that which was just as far from the truth. The eminent author who has written so much that is valuable in relation to the horse, had evidently given a good deal of time to observation and close study of the paces of the animal, and it was from no lack of penetration, but the absolute impossibility for the brain to record such active movements. His observations are worthy of being quoted, showing what was considered the intelligent explanation, before instantaneous photography had solved the problem. Stonehenge writes :

"To represent the gallop pictorially, in a perfectly correct manner, is almost impossible. At all events it has never yet been accomplished, the ordinary and received interpretation being altogether erroneous. When carefully watched, the horse in full gallop will be seen to extend himself very much, but not nearly to the length which is assigned to him by artists. To give the idea of high speed, the hind legs are thrust backward and the fore legs forward, in a most unnatural position, which, if it could be assumed in reality, would inevitably lead to a fall, and most probably to a broken back. It is somewhat difficult to obtain a good view of a horse at his best pace, without watching him through a race-glass at a distance of a quarter of a mile at least, for if the eye is nearer to him than this the passage of the body by it is so quick that no analysis can be made of the position of the several parts. But at the above distance it may be readily seen that the horse never assumes the attitude in which he is generally represented, of which an example is given at the beginning of this article. When the hind legs are thrown backwards, the fore feet are raised and more or less curled up under the knees, as it is manifest must be the case to enable them to be brought forward without raising the body from the ground. In the next act, as the hind feet are brought under the body, the fore legs are thrust straight before it; and so, whichever is chosen for the representation, the complete extension so generally adopted must be inaccurate. It may be said that this is meant to represent the moment when all the feet are in the air, and theoretically it is possible that there may be a time when all the feet are extended; because, as in the fast gallop the stride is twenty-four feet long, while the horse only measures sixteen from foot to foot, it follows that he must pass through eight feet without touching the ground, and during that time, as of necessity his legs must move faster than his body, the fore legs *may* change their position from the curled-up one described above to the extended one represented by all painters as proper to the gallop. Observation alone can therefore settle this question; but, as I before remarked, a race-glass at a distance of a quarter of a mile enables a careful observer to satisfy himself that our received ideas of the extended gallop are incorrect."

Following this is a picture of a horse with fore legs bent at the knee, and nearly parallel, and the hind legs thrust out behind. In number two of the instantaneous photographs, the hind and fore legs are all doubled up under the body by being bent at the knees and hocks. The near fore foot, which is nearly ready to leave the ground in number one, just touches the line which marks twelve inches, and that and the right hind foot are in proximity, the hind foot being a few inches in advance and the other being only a short distance behind. The right fore foot is much bent at the knee and pastern, and is the furthest from the ground, the hind feet a little more elevated than the left fore foot. The knee of the fore leg in advance and the hock of the near hind leg are about sixty-two inches apart, a striking difference from the ideal of the artist, or the position Stonehenge noted through the race-glass at a distance. The top of the withers and croup and the lower part of the muzzle touch the line representing the height.

"IN THE AIR."

The next picture, number three, the nose is on the line between spaces nine and ten, equalizing the two pictures as to the distance both should be. Three of the feet are still higher from the ground, and the back is very little above the higher horizontal line. The right fore knee is a little further forward, and that foot is not so high. The hind feet have been carried further forward, one in front of the other, and the right stifle is close to the abdomen.

Number four gives another new illustration, and would be thought highly absurd if the proof of its accuracy was not so emphatic. The right hind foot has struck the ground, the toe being nine inches in front of the line between the spaces eight and nine, eighty-five inches from where the fore foot was placed, and, consequently, showing this much of a flight through the air when all the feet were off the ground. The other hind foot is fifteen inches above the surface of the track, and in advance of the one which is on the ground, the right fore leg, from the knee down, being nearly vertical, the arm horizontal, while the other fore leg is much bent at knee and pastern, so that the sole of the foot is uppermost and level. Thus the

FIRST CONTACT

After the body is hurled through the air, is on one hind foot, and all our ideas of concussion and impinging force proven to be incorrect. This hind foot is thrust so far forward that the vertical line which strikes the ankle also strikes the cantle of the saddle, the foot being placed immediately under the rider, without much inclination of the pastern. This thrusting of the feet so far forward is to sustain the equilibrium as much as possible, and to give the greatest contractive force of the muscles.

In the next picture, number five, the same hind foot is on the ground, the pasterns so much bent that the fetlock touches : the other hind foot on the twelve-inch line, the fore feet more extended. Between the two illustrations, five and six, the near hind foot has come to the ground, and in six both hind feet touch. There is a space of thirty-eight inches between these feet, the right fore leg is extended to its extreme capacity, the heel touching the twelve-inch line, and the toe elevated so that the sole forms an angle with the horizontal line of forty-five degrees. The left fore leg is not so much bent at the knee as in the preceding two pictures, and the croup is three inches below the horizontal line which marks the height, some fifteen hands and an inch. Number seven shows the left hind foot and right fore foot on the ground, enclosing ninety inches of space between them, the other feet elevated and about one hundred and thirty-five inches apart. There is rather more sinking of the body below the line than in the preceding, with the head very nearly in the same position.

Figure number eight comes nearer the fancy of the artist than either of the others, and yet it is so different that it is still more conclusive testimony of the absurdity of the portraiture of the great limners of the horse. The right fore foot is on the ground, the leg only a trifle off the perpendicular, that little inclination being backward from the foot. The other three are extended, the right fore foot projecting in front of the nose, and elevated fifteen inches, the toe turned up. The left hind foot has just left its hold, and is four inches above where it rested. The right is raised sixteen inches, and at the very furthest extremity of its reach. Between the toes of those feet which are the furthest stretched apart are five and one-third spaces, equal to one hundred and forty-four inches; but this is aggravated, owing to the angle from the camera spreading and covering too much space on the screen. Before this was seen, an error arose in thinking that the hind foot had a retrograde action, and in the analysis of the stride on the back of the card are loops, showing this "back-action" movement. From the first Governor Stanford insisted that it could not be so, as all the movements of progression were forward, and closer reasoning established the truth of the position. In the Pictorial Gallery of English Race-Horses are pictures of Miss Letty, painted by F. C. Turner; Industry, A. Cooper, Royal Academician, and Atilla, C. Hancock the artist, the horses represented as galloping fast. All are after the stereotyped model, and if one fore foot was brought to the ground, with the leg perpendicular, they would be nearly in accordance with truth. But it is also just as sure that artists will stick to the old ideals for some time to come.

The pastern of the leg which supports the body is bent to a horizontal position, and in the next picture, number nine, the other fore foot is on the point of striking. This is the "leading leg;" the one from which the last bound was made, and when it reaches the ground the two imprints will mark the length of the stride. It requires two more representations, however, to give the full understanding. In number nine the body has been carried over the supporting leg until the foot is under the saddle, both hind feet are elevated to about the same height, and the croup is within two inches of the upper horizontal line. In number ten the left fore foot sustains the weight, having very nearly the same position as the right fore leg in number eight, but in place of the extension of the other fore leg, it has a backward inclination, the knee bended so that the back tendon is on the line which is twelve inches from the ground, the heel being above that line. The hind feet have become separated, though the hocks are near together, and the croup is a trifle higher than in the preceding picture. This is the position from which the cut of the last effort was made.

The last of the series I hold to be the most important of any. The eleven pictures give more than one stride, and consequently every portion of it is delineated. In number one the nose of the mare is marked by the line between seven and eight; in number eleven the line between seventeen and eighteen strikes her eye. This would give two hundred and seventy-eight inches to cover the stride of two hundred and sixty-five. More than this, the eleven cameras embrace a space of over five hundred inches, the horse being the centre figure in each picture. As was stated before, the toe of the left fore foot, in number one, was slightly in advance of the line between five and six, and in ten it is found on the line between fifteen and sixteen. In number eleven the foot is in the same place, but the body has been moved forward about thirty inches, and a perpendicular from the toe just touches the very back part of the saddle. The other fore leg has been brought forward so that the arm is at right-angles with the surface of the ground, the knee bent so much that the foot is considerably above it, in part owing to the bending of the pastern. All of that fore leg is above the twelve-inch line. The hind legs are drawn up, and though one hock is raised the highest, the feet both touch the horizontal line a foot from the ground, with no perceptible difference in the elevation. The position so fortunately caught by the camera at the right moment, is just as the body commences its flight through the air, and which will not be broken by contact with the earth until eighty-five inches have been covered. The last impulse of the hind legs to aid in this flight through the air was given one hundred and forty-two inches back, so that it is clear the great motor is in the fore extremities, and the old notion that the quarters and hind legs were the driving power, the fore legs only needed for props, to be rolled out of the way, is effectually exploded. It may appear tiresome to spend so much time on this feature of the action, though I think it will be found well worthy of our attention to give it proper consideration, and I shall be disappointed if I do not show, that in connection with other matters, it has a good deal to do with the explanations why weight on the feet is prejudicial to the race-horse, and, in some instances, an absolute essential for horses to trot fast.

Those who have the Muybridge photographs and have accompanied the description with a comparison of the pictures, will understand the action, but I despair of making it clear without their aid. Still, I trust that some idea can be derived of the true manner in which a horse gallops, aided by the three cuts which Gov. Stanford and Mr. Muybridge have kindly given me permission to use. Thus, the body is flung through the air for nearly one-third of the stride, the last and greatest force given by the fore-legs. While in the flight the legs are drawn up, hocks and knees bent, and the feet close together. There is a change in the position of the legs while the body is suspended in air, and one hind foot is thrust as far forward as it can be advanced, and the first contact is made with that foot. In a very short time, marked by a space of thirty-eight inches, the other hind foot comes to the aid of its fellow, and when the first that struck is raised a fore foot is down. Thus there have been both hind feet on the ground at one time, and then a fore and hind foot on opposite sides sustain the weight, though placed widely apart—ninety inches. The hind foot does not leave the ground until the fore foot is brought under the brisket, and then fifty-two inches in advance of that the other foot strikes, but evidently not until the first has left the ground. This is proven by the space. Notwithstanding the hind legs are so much longer, thirty-eight inches mark the step, and the photograph not only shows them on the ground at the same time, but from the placing, proving that both remain there for a period, uniting the strength of all the muscles. From the elbow to the ground an average-sized horse will measure thirty-seven inches, and after due allowance is made for the further extension which the straightening of the humerus affords, fifty-two inches is a greater space than can be covered by the fore feet remaining on the ground at the same time. Now comes into action the same foot which lent the last impulse at the commencement of the stride, and the strain on that is far greater than either the hind legs or the other fore leg has to bear. There has been none to ease it, or to bear a portion of the burden, and though the others have given the impetus, the last supreme effort is made by that. And made with a great proportion of the weight—at least two-thirds of that of horse and rider—to be raised at a disadvantage. Is it any wonder that horses "change their legs" so that the other may do its share of the labor? Is it surprising that there should be so many "bowed tendons" and "sprung sinews?" Under the teachings of the instantaneous photographs, is it at all remarkable that there should be twenty race-horses go wrong in front to one behind? We can now see what these great muscles are for which clothe the shoulder, and which are bunched up on the arm, and the powerful tendons attached to the knee, ankle and foot. The load has to be raised from this one leg two and a half times in every second, and it is this velocity which tells "the pace which kills."

But there is a far greater velocity which is not so well understood, and which is not so easily explained. If a horse is running at the rate of a mile in 1:40, and he strides two hundred and sixty-five inches, it is easy to determine the number of times each leg is moved. A timing-watch, tape-line, pencil, paper, and a little arithmetical knowledge are all that are necessary.

The position of the feet have been noted. We have seen that the fore foot left the ground when it was almost as far back as the loin, drawn up and thrown forward until struck again under the nose. The celerity of this movement is almost beyond calculation, and until Mr. Muybridge perfected his machinery there was a confused blur. The plates had to be made so "sensitive" that the shadow of a cannon-ball would be transfixed as it flew past, and Jove invoked to lend his lightning to open and close the slides which guarded the plates from contamination until the moment arrived, when the grotesque pictures could be produced with defined outlines. I will not venture at present on an estimate of the rapidity with which the feet are moved. That it is fast, all will admit who will give any thought to the subject.

Here is the key to the answer. In this lies the solution of the problem of the acceleration of the speed in the race-horse when shoes are replaced with plates—in this the vindication of the old proverb "that an ounce on the heel is worse than a pound on the back."

And now we will try an experiment: I take a cane, the length of the fore leg of a horse. It weighs eight ounces—one foot of the upper end balancing the other two feet. I move it through the air with ease, bringing it from a resting position into fast motion as quickly as I can, and I can stop it at the end of the circuit without expending a great deal of force. I reverse it, placing the heavy end on the floor. It not only requires a good deal more force to overcome the inertia, but more accordingly to stop it when it reaches the desired point. I can switch it without fatigue for some time—when clubbed it soon tires the muscles.

Fatigue has the effect to cause a sharper action of the muscles. When a horse is tiring, he commences to labor. The smooth, clean action is lost, and he "clambers," "sprawls," and then "goes all abroad." The horse when fresh has carried his rider, as the old huntsman expressed it, "as smooth as oil;" when "done," the loin bumps against the saddle, and the knees are raised with a spasmodic jerk.

When I come to the consideration of the effects of weight on the feet of trotters, I will give some illustrations which I have met in my own practice, and those which have come under my observation in that of others. It is needless to spend time in showing that race-horses can run faster with weight off their feet, as that is acknowledged by all whose experience is worth referring to; but the cause of it has been more or less a mystery. Heretofore I thought it was entirely owing to the change in action, and that a few ounces, more or less, on the feet of so powerful an animal as a horse could not lead him to such a difference. But now I firmly believe that even if the action were unchanged, the actual difference in weight would be detrimental, and that the ounce on the heel must have its effect. As I progress, there will be comparisons between the action of the race-horse and that of the trotter, and to avoid repetition we will now see what the photographs teach in relation to the trotting gait.

CHAPTER II.—ACTION OF THE FAST TROTTER.

More than two hundred years ago, the Duke of Newcastle described the trot in the following words, which are copied literally from his "New Method to Dress Horses:"

"*Secondly.* In a *Trott.* The Action of his Leggs, is, Two *Leggs* in the Ayre, and Two Leggs upon the *Ground*, at the same Time moved *Cross;* Fore and Hinder Leg *Cross;* which is the Motion of his Leggs a Swifter *Walk:* For, in a *Walk*, and a *Trott*, the Motion of the Horses *Legges* are all one, which his *Leggs* make *Cross*, Two in the Ayre Cross, and Two upon the *Ground Cross*, at the same time; Fore-Legg and Hinder-Legg *Cross;* and every *Remove* changes his Leggs; *Cross:* as those that were in the *Ayre Cross*, are now Set Down, and those that were upon the *Ground Cross*, are now pull'd up in the *Ayre Cross*. And this is the Just *Motion* of a Horse's *Leggs* in a *Trott.*" Making due allowance for the quaint form of expression, capitals and italics, the trot is accurately described and the diagonal motions correctly given.

It is safe to say the "Thrice Noble, High and Puissant PRINCE WILLIAM CAVENDISHE," as he is styled on the title page, never saw a fast trot, or he would have given a fuller description, as he dwells on the action of the gallop, describing it just as accurately as any of the latter day writers before instantaneous photography was known.

The first series of photographs—those which are mounted on cards and which were for sale by Mr. Muybridge—contained twelve representations. The cameras were placed twenty-one inches apart, with the vertical lines on the background twenty-one inches apart, and three horizontal lines, marking four, eight and twelve inches of elevation. In the later series the cameras were doubled, but as none of the latter have been offered for sale yet, I will confine the explanations to the first card. The subject was Abe Edginton, a horse which has a record of 2:23¾, and as square a trotter as can be found. He was probably trotting faster than a 2:24 gait when the pictures were taken, as Marvin sent him along at his best rate. In place of threads stretched across the track, wires were used, sunk below the ground, excepting a width of about two feet on the inner side, in which they were exposed. The sulky wheel, when it touched the wire, established the electrical circuit, and with so much greater uniformity than when threads were used, that there is more exactness, and each picture is a representation taken in front of the camera.

No. 1, of course, is the position he was in when opposite the first camera, and it was when the left fore foot and right hind foot were on the ground; the fore leg only a trifle from the perpendicular, the divergence being backward, and the hind leg well advanced under the body, with considerable bend in the hock. But inasmuch as there is an exact duplication in the stride of the square trotter, in fact two strides in what is universally called one, there is only a necessity for examining one-half of the figures on the chart. And in order to commence about the same period as in that of the race-horse, viz: when the "last effort" is made to hurl the body through the air, No. 3 will be

THE INITIAL.

In this, the line which divides the vertical spaces 10 and 11 is touched by the eye of the horse, as is also his lower jaw, the mouth being open, doubtless, owing to the tension on the bit. His right hind foot is in space 5, apparently about midway of the space which would bring it 115 inches back from his eye, and nearly at the extreme limit of its backward reach. The final propulsive spring has been made and it is ready to be lifted from its position. The fore foot on the same side of the body is as near as may be in the centre of space 9, the foot fourteen inches above the ground. There is a slight curve at the pastern, throwing the foot a little back, though the cannon is in a vertical line, and the knee is bent so that the arm is, as nearly as may be, in a horizontal position. The other fore foot, the near one, which left the ground only a short time before the hind, is in space 7, and is elevated ten inches. It is thrown back some two feet behind the elbow, and it is inside of the left hind leg, in juxtaposition to the ankle of the hind leg. This is the point when "scalping" of the coronet, or wounding of the ankle, shin, or hock occurs, and when the inquiry progresses further, this will be more clearly shown. The left hind foot is within six inches of the ground surface, and is about four feet in advance of the other hind foot. In the next picture all feet are

OFF THE GROUND.

Owing probably to the wire which was to open the slide of the camera, No. 4, being placed too far from No. 3, the representation covers a little more space than twenty-one inches. In all of the others the back of the driver and the eye of the horse are bisected by the lines between the spaces; in this they are a few inches in advance of the line.

The right hind foot, which was in the act of leaving the ground, twenty-four inches back, is now raised until it is fully sixteen inches up, and it is above the pastern some four inches. The pastern is so much bent that the sole of the foot is turned up, but from the pasterns to the stifle the front part of the leg forms a straight line. The right fore foot is thrust forward so that there is only a slight bend at the knee, and the toe is within eight inches' of the surface. The feet, on the same side of the body, are about ten feet apart, the front being across the line between spaces 11 and 12, the hind in the centre of space 6. The left fore foot is in space 9, twenty inches from the ground, four inches higher than the knee, the cannon is horizontal, the arm forming an angle rather more acute than a right angle. The left hind foot is within four inches of the ground and under the elevated fore foot. This figure is nearer the artist's conception of the fast trotter in motion than any of the others, but the next is a startling innovation on previous opinions.

The first picture of a trotter at speed, which was engraved from a photograph taken by Muybridge, was that of Occident (see cut), and the fifth of this series is nearly the same. In that of Occident the feet had come in contact with the ground; in this the hind foot is almost touching, the fore being a trifle higher. At first sight it does not convey the least idea of speed, or even motion, and when first published elicited ridicule from nearly every one who fancied they had a perfect knowledge of the horse. Had those of the gallop preceded the trot, there would have been still more exuberant jollity, but by the time they made their appearance people had given some thought to the subject, and though many are still sceptical, a man of any intelligence who has examined the representations, must acknowledge their truth.

While the feet on opposite sides touch the ground at so nearly the same time that the sounds of the contact is merged into one in the square trotter, the hind foot probably touches first. The other hind foot is thrust backward five feet behind it, and is elevated fourteen inches. The raised fore foot nearly touches the breast, the knee folded so as to make an acute angle of the arm and cannon, the bending of the pastern giving the foot the greater elevation. In considering the effects of the weights, this is one of the most important positions to study, as the flexors have performed their duty, and from thence the contra-acting muscle will carry it forward. The sixth figure on the card—the fourth which I have taken to illustrate the stride—is nearly the same as the first, excepting, of course, the reversal of the feet. The feet on the ground are firmly planted, the body having been carried over them until the front leg is nearly perpendicular from the ankle up; the pastern is inclined backwards until it is nearly horizontal.

The hind pastern is still more bent, so that the ankle is lower than the coronet, and as the foot is under the loin, the hock has to be bent in order to advance the foot to the furthest point. The other hind leg is in an apparently awkward position. From the stifle to the hock it is nearly parallel with the body, and the cannon is vertical, the pastern being still crooked. This foot is on its forward journey, having been carried from space 7 to space 9.

THE RAISED FORE FOOT

Is still nearly as high as in the preceding figure, and though it has not been carried as far proportionately forward as the hind, it is now on a line with the arm of the other leg. The sole of the foot is close to the arm of the same leg, the arm level, the bending of the knee and pastern sharp.

In the next picture, the body having been carried twenty-one inches forward, the feet on the ground are placed that much back relatively. This has straightened the pastern of the fore leg somewhat, that of the hind being still down. The fore leg is now exerting its final impulse ere it leaves the ground, and the straightening of the pastern of the hind leg will complete the application of the force. The non-motors, during this period of the stride, viz: the left fore and right hind foot have been advanced, the former is on the line which divides spaces 13 and 14, and the other on the line between 10 and 11. The fore is raised to a height of eighteen inches, and is two feet in front of the other fore foot, so that it must have moved rapidly during this short period; and this motion, if I am not greatly in error, has a bearing on the question of the effects of weight, which, however, will come in more appropriately hereafter. The hind foot is advanced until it is parallel with the other hind foot, and elevated some distance, so that the hock of that leg shows above.

The next figure is nearly a duplicate of No. 3 on the card, the only difference being that scarcely half of the stride has been made. By actual measurement from between the prints of the left fore foot the stride was $18\frac{1}{2}$ feet—222 inches. The eye of the horse was on the line between 10 and 11 in No. 3, and in figure No. 8 it is on the line between 15 and 16, five spaces, equal to 105 inches. Six inches further would give an exact duplicate.

A lesson from this is gained. In No. 3 the right fore and left hind feet were together; in No. 8 they are at least six inches apart, showing that at this portion of the stride the hind foot is in rapid motion forwards, outside of the general progression, while the fore foot is moving slowly, that movement being principally upwards.

I desire to call particular attention to this fact, that when the fore foot first leaves the ground it is elevated more slowly than the hind advances, not giving room for the hind foot and leg to get in a proper position; and in some cases it is so dilatory that the front part of the hind foot, and the shoe or sole of the front are brought in contact; a blow on the horn causes pain, or the coronet is wounded. If the hind foot is carried wide enough to escape, the ankle shin or hock receives the injury, and in either case the animal endeavors to remedy the trouble by a change in the action. Unfortunately it has not progressed far enough to understand that this can be accomplished by quickening the motion of the fore feet; and knowing that a run or canter does not entail the injury, the thoughts are first directed to relief in that way. But from having been punished for forsaking the trot, a compromise is tried, and "running behind," "single footing," etc., is resorted to. Before entering into this portion of the argument, it may be as well to review the trotting action as exemplified by a horse which can show that action properly, and is fast enough to display the peculiarities. The slow trot, such as the Duke of Newcastle described, is entirely different from that which will carry a horse a mile in 2:20 ; though, of course, there are points of similarity. How to perfect that slow natural movement into the flying trot is the important question, and fortunately Mr. Muybridge has given a series of eight pictures representing the same horse, Abe Edgington, "jogging" at an 8-minute gait. Figures 3 and 6 of that series show very plainly the danger of injury at even this slow rate, and explain why young horses endeavor to avoid the trouble by "hitching," etc., at the outset of their education.

In No. 3 the left hind foot is on the point of striking the ground directly under the right fore foot, the toe of which so nearly touches the coronet that there is no separation visible. In No. 6 the other hind foot is coming down, and the left fore foot has the toe still resting, with the bottom of the shoe vertical, and in close proximity to the wall of the hind foot. These two figures are the nearest approach to the feet being all off the ground, and there is very little elevation of the hind feet during any part of the stride. The fore foot, at its highest position, is only a little over twelve inches from the ground, and the forward motion is obtained with very little bending of the knee in comparison with that which marks the faster pace. There is also a great lack of energy, as the body is in the air for so short a period that a few inches measure the space, and the hind foot is slightly, if at all, advanced beyond the imprint of the fore. The stride is about eleven feet, so that it is evident that the great acceleration of the speed is due more to the rapidity of action than to the ground covered. "Jogging" is an expressive characterization of this gait, as it calls into service only a small portion of the muscular force of the animal. This is the groundwork, however, to build the superstructure upon ; and the slow gait must be studied in order to discover what is necessary to perfect it into the faster. The student will perceive that the greatest trouble is in the dilatoriness of the front feet ; and without the photographs he will discover that the hind feet are brought into close proximity with the fore. Many horses "forge," strike the toe of the hind shoe against the web of the front, and in spite of the art of the shoeing smith or the skill of the equestrian, persist in the habit. Something may be effected by "throwing them more on their haunches," and compelling cleaner action by a tight rein and the application of the spur; though many remain incorrigible under the best hands, and as soon as the momentary fear of punishment passes away they relapse into their customary slovenly paces. Horses which have naturally a long stride are more prone to exhibit this style of trotting than those which are more prompt, and hence the Morgans were favorites on account of the spirited manner in which they handled their feet. This forging takes place soon after the fore foot has left the ground, and immediately before the hind foot is brought down. If the front foot can be got out of the way the remedy is at hand; and, without dwelling longer on the phases of action, excepting as they are presented by the consideration of artificial appliances, I will proceed to that part of the subject. Without a full series of the pictorial representations of horses in motion, I despair of making it entirely intelligible, though I think enough has been shown to base views upon which will be found logical.

CHAPTER III.—Action Controlled by Artificial Appliances.

It is oftentimes troublesome to give a clear explanation of a subject with which one is familiar. There are questions which test the scrutinizing faculties to the utmost capacity, and which, after years of study and many experiments, are still in doubt. Simple as it may appear to those who have not endeavored to unravel the mystery, the reasons for artificial appliances controlling the action of horses are not clearly apparent, and conjecture has to play a prominent part in the argument. A good deal of the reasoning is forced to be hypothetical, depending on analogy, and trusting to laws which govern in cases which are somewhat similar. Not the least perplexing of these questions is the effect of weight on the feet of horses, especially the power of governing trotting action by the application of weight on particular parts of the foot. The latest inventions of compact, metallic masses, placed on the outside of the hoof, have effected wonderful changes, and in so many instances that the advantages are beyond cavil; and the trainer who does not avail himself of the benefits to be derived from a proper use of them, is far behind the requirements of the age. From their use fast trotters have been multiplied, and speed and steadiness resulted. There are so many illustrations of the benefits which have followed their adaptation, that there is little necessity for rehearsing the numerous instances in which they have proven successful. Every observing trainer of trotters is aware of what has been accomplished; every man who has paid any attention to the trotters of the last ten years, must concede the wonderful effects, if he has pursued the investigation with an ordinary amount of intelligence. Trainers, breeders, and fanciers of the trotting-horse have witnessed injuries resulting, as well as good offices; and before the application of the weight was properly understood, it was thought the drawbacks were more to be feared than the advantages justified. These have been overcome by a better understanding, and the knowledge which years of experience have brought has reduced the casualties to a minimum. In all probability, there are many instances where the use of weights has saved the animal which wore them, by doing away with the necessity of harsher methods, and by lessening the strain which would follow harder work.

I shall take the ground that the good results which follow the use of toe-weights are dependent, first, on mental influences; secondly, on mechanical effects. The former I hold to as by far the most potential of the two, and the grand secret of the success which has resulted from their judicious application. It may be better, perhaps, to state that the effect of toe-weights is to fix a different habit of action in the mind, the intelligence of the pupil being called into requisition, and this is supplemented by mechanical laws which aid in the development. I am aware that this position will be thought untenable by nearly every one who has studied the subject, though I feel the utmost confidence in the correctness of the statement, and will present the arguments with a degree of positiveness which has arisen from long study and careful examinations. I was forced to this conclusion from absolute necessity. That there was some particular configuration which required weight on the feet to counteract the wrong formation, was the first idea which came to my mind, and from the pacing gait being so susceptible of a change to the fast trot, there was some foundation to commence upon. From many pacers having a form which differed from that which is usually seen in good horses, we are prone to associate these peculiarities with the pacing gait. A sloping quarter, the stifles placed low, a good deal of bend in the hock, and with high withers, narrow chest, and fore-legs close together, is called the pacing form. But the very fastest pacers I have seen were of a different pattern. Longfellow, Jim Brown, Nimrod and Hiram Tracy are straighter in the quarters than a majority of trotters, and though Defiance, Lady St. Clair and Ben Butler have a good deal of angle in the innominatum, it is not enough to attract particular notice. Those were all fast pacers, and Defiance has the fastest record at both gaits, having paced in $2:17\frac{3}{4}$ and trotted in 2:24.

There was a pacer owned in Chicago which had a pacing record of 2:30, though he could go very much faster in a brush than that rate. He was driven by the side of Prairie Boy, a trotter with a record of 2:32. The owner of the team had a farm some twelve miles from the city, and, returning from thence the same day he drove there, the horse was wearied. Part of the distance there was deep sand, which increased the weariness of the pacer, and, before he got home, he changed his gait to a trot. This was the first instance when he had shown any capacity to leave his natural method of progression, and his driver was a good deal astonished at the change. The next day he drove him and his mate to the track—the Chicago Driving Park—and, much to the surprise of every one who witnessed the performance, he not only trotted, but could carry Prairie Boy off his feet and force him to a sharp run. After that display, he was bought by some parties and put in training, and there was not the least difficulty in making him trot. It was claimed that he showed a quarter in thirty-one seconds, and it was certain that there was not a horse on the track which he had the least trouble in beating a long way off through the stretch. From some internal trouble, he could not keep up his rate further than half a mile, and was entirely useless for track purposes. This was before the era of toe-weights, and there was no change from the shoes he wore on the road. In those same shoes he paced until he became wearied from the journey, and the deep sand completed the lesson.

In this case there was a preternatural adaptation to benefit at once from instructions—something of the same character which enables Blind Tom to play a difficult piece from hearing it once—and is such an unusual occurrence that it may never be witnessed again. Still, it has a bearing on the question, for it is palpable that there was nothing in the way of this horse trotting fast excepting the knowledge of the proper action which was requisite to enable him to do so, and this knowledge came as nearly intuitively as it could. There was no "balancing" necessary, no change of centre required; the machinery was there, ready to run whenever the brain took the direction. The case of Defiance presents some similarity to the one recited. He had paced in races for years, gaining his record of $2:17\frac{3}{4}$

in 1873, and in 1872 he made a dead heat of two miles with Longfellow in 4:47¾. In 1874 he paced a number of races, and was taken East, returning in the Fall.

After his return he was placed in training to trot, and from the first he could trot very fast, but owing to his long schooling at the pacing gait, it was a troublesome affair for his teacher to overcome the settled propensity. He was weighted quite heavily, and when not permitted to pace he would take a hand gallop, and appeared to have come to a determination to be as annoying as possible. His trainer was a man of a great deal of experience, of an equable temperament, and with confidence enough in himself to pursue the course his judgment dictated. After patiently endeavoring to educate him by the gentle method, he adopted the plan of running him until he became so tired that he could neither run nor pace, and by that means he succeeded in educating him to the desired point. The weight made racing very fatiguing, and it was inimical to the lateral motion; and when it came to the choice between the gallop and the trot, the latter was adopted. As early in 1875 as April 29th, he trotted a race against four competitors, winning in "straight heats" in 2:38, 2:34¼, 2:35¼. On May 1st he was again victorious, in the improved time of 2:28¾, 2:30¼, 2.27¼; and on May 11th, he trotted against Abe Edginton, in a match for ten thousand dollars a side. It was a very remarkable race, in which it took five closely-contested heats to decide it. Defiance won the first in 2:24¼; the second, in a jog, in 2:29, and retained the lead in the last until only a short distance from the goal. Every heat was fought for inch by inch, and owing to the track being soft on the inside, a great deal of ground was lost on the turns by the horses being driven on the extreme outside, which added at least two seconds to the time of each heat. In one of the heats he lost, Defiance broke not very far from the score, and finished on a pace, as much probably owing to his driver not caring to make an effort to recover him in the few yards between the mishap and the wire. In the scoring for the five heats, he only came up once on the pace, and at the finish, amid the yells of thousands of spectators, he was as constant as a clock. The finish of the fifth heat is thus described in the account published at the time:

"Both are now fairly in the straight work, and the grey is gaining. He is surely gaining, yet there are no shouts from enthusiastic partisans. It is too exciting for cheers. Half-way home Edginton leads; he is half a length in advance; he is increasing his advantage. Muscles, lungs, heart, brain, all are tired in the bay. You can hear his sobs as he approaches the outcome. He reels; his ears droop; his eye is bloodshot as he staggers over the score a length behind Edginton, who has won in 2:26." This was a surprising race for a horse to make whose trotting education had only occupied a few months; but there were other drawbacks to contend against. The ankles of the hind legs were all "stove up," double the natural size, and so badly "cocked" that the weight was thrown on the toe to an undue degree. Notwithstanding these ailments and the hard race with Edginton, the day following he took part in the 2:27 race against Sisson Girl, Ajax and George Treat. Strong elastic stockings were worn on his ankles, binding them so rigidly that many thought they were an injury. He won the first heat in 2:26$\frac{1}{4}$, in the teeth of a heavy gale, and led in the second to beyond the half-mile pole, when he made a break, and refused to trot again until he was a double distance out. In the match against Edginton, there was 196 pounds in the sulky, and 190 pounds in all of the other races. On December 11, 1875, he trotted against St. James. When driven by Hickok he scored 2:27—2:30—2:24. All this was accomplished within a year after the commencement of his training to trot, and after he had been pacing for many seasons.

These two "conversions" from pacing to trotting are worthy of consideration. The first case shows that the horse had happened on a manner of going which was more satisfactory than pacing, and he had the faculty of retaining it under pressure. All that was necessary was to move the bit in his mouth if he attempted to start on a pace, and when once he struck the trot it was almost impossible to force him out of it. As has been stated before, when the animal becomes tired there is more knee-action—a sort of spasmodic jerk to overcome the sluggishness of the tired muscles. The sandy road increased the tendency to a sharper bending of the knee, and the two combined had the same result as the application of weight on the toe.

In the training of Defiance, the toe-weights, in a great measure, did away with the desire to pace, and when he broke from the trot his inclination was to indulge in a hand-gallop. When forced to run, the toe-weight is a terrible incumbrance, and when the "last effort" is made, to hurl the body through the air, there is an immense strain on the muscles and tendons of the fore leg which gives the final impulse. The whole weight of the body has to be flung to a distance, and though a portion of the impetus has been derived from the other legs, by far the greatest part of the force is due to the motion which precedes the bound. To raise the weight on the toe there has to be a violent movement at the instant it leaves the ground, and then it has to be raised to an altitude of eighteen inches and thrown forward with a velocity which Muybridge estimates to be a hundred feet in a second, or double that of the body.

Whoever will give proper study to this phase of the action of the race-horse, will readily understand the difficulty there is in overcoming the inertia of even a few ounces of weight on the toe; and, as it is well known, that the application of weight induces higher action, there is further loss of power arising from the exuberance of muscular effort. The old-time quarter-horse men were well aware of the effect of unequal weight on the feet, and a favorite plan to mislead those they desired to hoodwink, was to "cross shoe" the horse, and then arrange it so as to run a night trial, when the trick could not be discovered. Cross-shoeing consisted in putting a heavy shoe on a fore and hind foot on opposite sides, and a light one on the others, and this would make a difference of many yards in a quarter of a mile. In some cases which came to my knowledge, an arrangement was made to leave the stable unguarded, when a confederate would suggest stealing the horse out for a surreptitious trial run, after the change in shoes had been made, and the result would be that the capacity of the animal was underrated and the desired match obtained. In so short a run as a quarter of a mile the difference between shoes and plates is not so great, and weight on the feet would not be so effective, but the inequality broke the regularity of stride, as well as the drawback of additional weight, and the purpose was thoroughly effected.

Defiance soon learned that running was an extremely toilsome manner of progression with a pound weight on the toe of each fore foot, and he also realized that it was inimical to the pacing gait. But the severe exertion of the five tremendous heats with Edginton had made him so "sore," that when called upon to trot, the following day, he resorted to the hand-gallop to escape the punishment the fast trotting inflicted. I have read an account of a plan adopted by Astley, of circus renown, that is pertinent to the question of mental influences on the horse, and which is so well authenticated that it may be received as the truth. A horse, after a due course of education of the period, consisting mainly in severity, performed his part satisfactorily; but after a short time, though faithful at rehearsals, in the presence of the assemblage he balked. The unmerciful castigation could not be inflicted in the course of the play, and he took the advantage of his knowledge that he could shirk his task with impunity. The plan was adopted of filling the circus with spectators, and when, relying on their presence, he refused to go through with his part, the punishment was inflicted with extraordinary severity. This destroyed his confidence in the immunity which a crowd of people afforded, and there was no more trouble with him. A still better illustration, and one which is more nearly analagous to the subject under consideration, I witnessed in Chicago, in 1857. At that time Franconi was there, and the trick-horse he brought from France was ailing. He bought a roan horse called Chicago Top, and I took a great deal of interest in witnessing his manner of training him. Part of his system was weights applied to the ankles when teaching him to dance, and these were supplemented, at times, with strings of small bells. The effect on his action was astonishing. Knees and hocks were sharply bent, and this greater elevation of the feet and increased rapidity of stroke enhanced his capacity of keeping time to the music. When performing, the bells were kept on, and this was supposed to be for the same purpose of castanets in the cachuca and cracovienne, but it was in reality a reminder of the lessons he had received.

I have known the small bells successfully used in training trotters, and the old-fashioned "rattles" owed a portion of their efficiency to the rattling of the beads. These were spheres of lignum vitæ or bone, strung on a strap which was buckled around the pastern loosely, and the effort to throw them off caused a sharper bending of the knee. There are still cases where bells or rattles will be found beneficial, notwithstanding the modern appliances have almost banished them from the trotting stables. Many of the younger trainers never saw them, and would be at loss to know what they were intended for.

In fact, the information in the "Dexter circular," with which the country was flooded some fifteen years ago, and of which the sales were enormous, was merely a recommendation to use the rattles, which would be sent on receipt of so much money. The cost of the circular was $1.00, and a great number who got them supposed it was a new invention. In the preliminary newspaper advertisements the claim was made that a great discovery had been made in the education of trotters, with the usual number of certificates, guaranteeing its efficacy. After all, it may be that a majority of those who obtained the information were benefited more than the cost of the circular and rattles.

CHAPTER IV.—Unsolved Problems.

When I wrote the preceding chapters of the appendix, about two years ago, I then thought I had a fair knowledge of weight on the feet of horses. Now, after that length of time of study, observation, and the tests of many experiments, I am in doubt, and in place of offering rules for the guidance of others, am forced to admit a want of confidence to make statements with any degree of authority. Further than that, I have the same lack of confidence in the opinions of others, and incline to believe that the future must be depended upon for satisfactory elucidation. There has been a wonderful improvement in the manner of applying weights from the rude contrivances at first in vogue; there has been a great increase in the intelligence which has led to a more rational use, and yet there is a void as annoying as it is puzzling.

That fast trotters have been made by the use of weights is beyond even the cavilings of the hypercritical; that many promising horses have been ruined by the abuse of them is equally true. For a time there was a mania, an infatuation which became epidemic, and from Maine to the Pacific every track had a majority of horses encumbered with these appendages. Probably Indiana, for the number of trotters in training, gave them the greatest prominence. Saddle-horses had been prized there ever since the settlement of the country, and the popular saddle-gaits were cultivated. Kentucky and others of the Southern States were equally as anxious to improve the horses that played so important a part in daily life, but with the exception of a portion of Tennessee, and some sections of Missouri, Kentucky was the most indefatigable in perfecting and breeding the trotter. The pacer and those so often confounded with the pacer, viz., the rackers, were more readily "converted" by the use of weights than by any other method.

This was also the case with broken-gaited trotters, and hence whenever a horse was of these kinds he was soon entered in a course, in which the most potent educator was weight on the feet.

I have stated, in the previous chapters, that my belief was that the mental effects were greater than the mechanical, and a few illustrations were given to sustain the argument. I still think that position is correct, and though the mechanical is the first effect, after that comes the benefit derived from the reasoning faculties being made subservient. The animal has been compelled into action that is more favorable for speed; his intelligence then comes to his aid, and, cognizant of having mastered the difficulty, he repeats the method of handling his feet and legs when the monitor is removed.

There is another phase of the toe-weight problem that adds greatly to the trouble in arriving at a correct solution. While broken-gaited horses are generally amenable to the good effects of weight, in a few instances it has been found impossible to correct the faulty action with weights, and then, perhaps, some other contrivance may work a cure. Again, there are two horses apparently identical in their gait. Weight aids one of them, the other it makes worse. Twelve ounces may be required to accomplish in one what four ounces may do in another, and so the paradoxes apparently come into notice at every stride in the journey; and, loth though we are to acknowledge the ignorance, it is so palpable that it cannot be denied.

With all this ignorance there rests the information that certain valuable results have followed the use of weights; and, though I cannot go so far as my friend H. D. McKinney, of Janesville, Wisconsin, in saying that "I would never try to develop a colt if I could not avail myself of the use of toe-weights if I needed them," I should be at a loss what to do in many instances if debarred from their aid. And, by the way, I have always considered that Mr. McKinney was "better posted" on weights—the various kinds, uses and abuses—than any other person in the country. He invented some of the best patterns of their time, and was largely engaged in the manufacture. Being a practical horseman, skillful in breeding, rearing and driving trotters, and a man of education and intelligence, it could not be otherwise than that he must obtain an esoteric knowledge, valuable as it was complete. He wrote for the New York *Sportsman* an essay which was published in that paper April 30, 1881. That essay contained a good deal of sound logic, and yet I imagine that he was troubled somewhat as I am, and did not feel very positive in giving reasons why the results he instanced followed the use. For instance, he credits weight with giving a tendency to "sharper folding of the knee," and reasons that St. Julien's manner of progression indicated that weight would increase his speed, and Maud S would do better without them: the reasoning, of course, being from a theoretical point of view. He also states that "many horses that cut their elbows when shod with an ordinary pound shoe have been prevented from doing so by using an eight-ounce shoe and four-ounce weight." Now, as weight on the wall, especially if placed high, is credited with giving sharper knee action, in that case the elbow should be struck still harder. But practice proves that this is not the case, and we are all aware how superior are the teachings of that over the most plausibly constructed theory in the world. People, however, are becoming convinced that much bending of the knee is not so essential as it was deemed a few years ago, and hence there is not the necessity for appliances to give that excess of motion. Then, as in the case of the "knee-knocker," it has been found that this exuberance may be modified with the application of weight, and recently John A. Goldsmith informed me that he intended to apply them on the Santa-Claus-Sweetness two-year-old to "carry him out." This young trainer has been remarkably successful, both in the use of weights and also in discarding them when the proper time came. Director is an instance of the latter, and after having been deemed one of the sort that required heavy weights to balance him, he can now trot very fast barefooted. In a conversation with O. A. Hickok a few days ago, he told me that he saw Director move around the first turn of the Bay District Course in $36\frac{1}{2}$ seconds without anything on his feet, and he went with great ease and fine action.

In the essay alluded to Mr. McKinney gives many instances of the efficacy of toe and side-weights, and, doubtless, in the two seasons since he wrote, many others have come under his observation. It is not proper to take the views that were expressed even so short a time ago as two years, when the subject is so provokingly mutatious as that under consideration. First, it was thought that masses of steel in the shoe and correspondingly huge appendages in the weight were an absolute requirement. Then there have been changes in everything pertaining, and from the crude fastener of a spike between the shoe and the foot to the latest patent, there is no wider divergence than in the reasoning all have been instrumental in evoking. I looked for a key in the instantaneous photographs of horses in motion, and as these gave the first correct knowledge of the action of a race-horse, was sanguine enough to think that with their aid the problem would no longer mystify. I pored over and studied the first cards that were published, and when Mr. Muybridge sent me a set arranged for the zoetrope, I was so eager to go at the lessons that I made an instrument which proved as good as need be to reproduce the exact motions of life. There has scarcely been a night that I have not set the wheel revolving, intent on discovery. The gallop, trot, pace, single-footing, gave plenty of chance for comparison; and still, so far, I am in a labyrinth with one thread, perhaps, to drop in the journey. In a fast, square trot there is more danger from the fore and hind feet coming together in what is called scalping than at any other gait. It may be the only gait in which this can occur, as trotters and gallopers are the only horses in training I have had for a number of years, at least twenty, and the representations are my only guide. When a colt commences to hitch it so oftentimes is the cause that there will be almost a certainty of finding that to be the trouble. Scalping-boots and speedy-cuts do not always correct the difficulty, as a touch on the leather is a reminder of the pain of previous blows. Being aware that the injury is not done in a gallop, an approximation to that gait is depended upon for relief, and it takes a long time before the fear is overcome. The toe-weights give an accelerated movement, and when the hind foot passes under the front, it is so far elevated as to give room without danger of collision. By referring to the cuts of the trotter on pages xviii and xix it will be noticed that the two fore feet are very close together when the last impulse is given to send the body through the air, and when the hind foot gives the last nervous push and is elevated a trifle, the toe of the fore-foot on the same side is in close proximity to the hind-foot, which is thrust partly under it, if even they go clear.

The sharp bend in the knee, as shown on page xxi, is when the supporting legs are nearly vertical. This part of the stride is when the legs are so far apart that injury is impossible excepting to the elbow, and the scalping wound is given when the fore-foot has just left the ground. To overcome the inertia of the toe-weight, there is a quicker motion—a sort of snatch—which takes it out of the way before the hind-feet come under the front. It may require quite a heavy weight at first to call forth the muscular energy necessary, and as the animal becomes accustomed to this action, the weights are reduced, and, perhaps, finally abandoned. As has been stated, Director is an instance, and Mr. McKinney presents another, and I give it in his own words: "Last season I purchased a young mare at a round price. She was wearing pound shoes and six-ounce toe-weights. I thought her fast, and paid for speed when I bought her. I sent her to the track; four weeks did not improve her speed. I brought her home, and took her in hand myself. First time I speeded her I saw the toe-weights were not wanted—made her fold too much, and caused her to dwell—and they were dispensed with. Next day I concluded the shoes, instead of weighing a pound, should weigh no more than twelve ounces; in fact, removed eleven ounces from her feet, and in ten days I improved her speed ten seconds. Now, the man I bought her of was all right in weighting her at first, as she lacked action when he bought her, and the weights made her quite a trotter, but he did not know when to begin to unload."

There are so many illustrations of the same kind that it would require chapters to give a small portion of them, and any person who has the least acquaintance with horses that have worn them is aware of the fact. Now, if it were a mere mechanical effect akin to the counterbalance on a wheel, to which a crank and piston is attached, the weight would be an imperative necessity at all times. If even the effect were due to the position the weight bore to the extensors, as has been claimed, there it would have to stay. But granting that the pupil realizes that it has learned a better method of handling its limbs to progress according to the wishes of the driver, then it is easily understood why the weight can be thrown aside when the lessons have been learned. Even with the larger intelligence of the human family, mechanical contrivances are resorted to to correct faults in the "style" of the athlete, the dancer and the soldier. The proper swing, as it may be termed, cannot be mastered without using methods that compel the proper action, and these have to be persisted in until habit overcomes the dilatoriness of natural gifts. When the jerk was first given to overcome the unusual weight the animal discovered that the usual injury did not follow. Thousands of times repeated, confidence was established, and by that time it became much easier to make the proper motions. With that result the weights have accomplished the purpose, and further than that, there is the drawback of carrying the "ounce on the heel."

In the first chapter of this appendix I wrote that weight is a drawback to the race-horse, an advantage to the trotter. I will have to modify that statement after the lapse of two years, and change it so as to read that weight is advantageous in the education of many trotters, but I am greatly in doubt of any benefit accruing after the schooling has progressed to a certain stage. Further than that, I have faith that the trotter of the future will be relieved from the incumbrance of heavy masses of metal on the feet, whether in the shape of shoes or weights, though the latter may always prove one of the main resources for the correction of wrong movements. It is rather mortifying, however, to surrender a position thought to be so well fortified that the fort could be held for any length of time, and in place of an essay replete with erudition, admit an ignorance that I am well aware exists. To be candid, I must confess that to reconcile the conflicting testimony is beyond my capacity at present, so far as an explanatory elucidation of the reasons for well-established effects, and in lieu of such a dissertation, will give a description of some of the kinds that have come under my observation, the reasons I have for a preference for some of the patterns, and the difference required in those that are worn with tips. This will include some novelties confined to my individual practice, and which, so far as tested, are more satisfactory. It will, of course, be unnecessary to occupy space with those patterns which have been generally discarded, and scarcely proper to select favorites further than to indorse what I consider the most correct plan of applying the weights, and the principles underlying a safe use.

The most convenient method of attaching toe-weights is by using a spur, which is welded to the front part of the shoe or tip. The convenience arises from there being no trouble with other kinds of fastenings, as in one patented invention the only thing necessary is to slip the weight on, when the bevel of the spur and the tapering form securely holds it in place. It is evident that every time the foot strikes the ground the more firm will be the clasp, and to release it several sharp blows with a hammer will be necessary. With this, of course, there can be no change in position, and to give the power to regulate the height of the weight on the foot, a screw has been added to some of them. A thread is cut in a hole in the weight, and the screw acting in this, presses against the spur, and any desired elevation can be obtained.

But the great drawback to the stationary and unremovable spur is that in the first place it is unsightly, and a still graver objection is that there is a chance for the horse to injure itself. Very frequently a horse will scrape a foot against the opposite leg, and in that case the sharp edge of the bevel will do damage. When the animal is walking, led by the bridle, a sudden start from fright, or, perhaps, in play, the legs are crossed, and injury follows.

Last Summer I saw a mare belonging to Mr. Titus which caught the spur and bent it so far forward that it was in a horizontal position when she came back to the stable. Nearly all the appliances to do away with the objections of a stationary spur have made necessary the mutilating of the horn, in order to make a slot through which the part that fastened it to the shoe could be thrust.

When the spur is detached there is an ugly looking cut in which gravel and clay will accumulate, and, in some cases, become so firmly impacted that it is troublesome to remove. In the many experiments I have tried with weights there was one plan that I thought highly of until I fell on one much simpler, and in every respect better, and which will be described hereafter. In that which has and that after a few times the thread in the horn will be worn smooth. That it is effective I know from the tenacity with which the live horn will hold even a smaller screw than those which are used ; and if the screw is driven home there is not motion enough to cause it to lose its hold. Those that I have tried have demonstrated this view to be correct, inasmuch as on two occasions the quarter-boot slipped, and the screw still held the weight in place. Although I have used mine for some time before date of the patent which covers the fastening with screws to the horn, it is probably an infringement, and in describing it it must not be considered a warrant for others to use it. I took a piece of sheet brass, an eighth of an inch in thickness, and cut it into the form I wanted. It is something of the shape of an oval with a continuation below to the desired length. If the design was to have the weight low on the foot it was short, and if high, long. Near the bottom a hole is made for the screw to go through, the corresponding hole in the horn being from a quarter to half an inch above the junction of the horn and tip. The desired weight, made of the same sheet brass or thicker copper is riveted to the oval-shaped part, having the same concavity to fit the convexity of the hoof. The part next the foot is carried up nearly to the coronet, and an outside strip is also made of the same length. The object of this is to form a slot through which the lower strap of the quarter-boot is passed, and the upper strap is run through a leather loop constructed as follows : Being impressed with the idea that there should be some elastic material between the weight and the foot, a piece of leather of the same shape, though covering rather more surface, is used for lining, the rivets which fasten the plates together holding it, and the upper part is turned over back to the rivet, forming the loop. This is to have a softer material near the coronet than the metal. The quarter-boot must be made with double straps and to cover a part of the heel so as to insure it from slipping down. The straps are buckled when the weight is higher on the foot than it is to be worn, and then pushed down so that the screw can be introduced. Notwithstanding the added thickness of the leather lining the longest screw I have used has been five-eighths of an inch, No. 9, and in most instances one-half an inch of the same number. Before using the screws in the horn, and previous to the keystone fastener being thought of, I tried cutting a thread in the tip and the horn, half in each. But the thread on the metal part was too fine for the horn, and it was unsatisfactory. At that time I did not think of using the ordinary wood-screw, and the small taps and dies that I had were very finely cut.

The use of the wood-screw not only taught me a better manner of fastening the toe-weights, as it also helped me out of a difficulty in attaching scalping-boots to a bare foot. To keep these in place I thought it was imperative to have at least one projection at the heel in order to fasten it properly, and in the case of Anteeo, followed the practice, heretofore described, of wearing a three-quarter shoe. With that I used the keystone at first, and then the wood-screw driven into the horn. A thin piece of metal was let into the quarter-boot for the screw to pass through, and I found the method far superior to a strap drawn through a slot. But Antevolo also needed the protection of a scalper, and every plan I tried of attaching it to the speedy-cut above was a failure. I tried several ways without any success, and as a last resort put on the boot the same as if there was a projecting heel. Luckily the scalper was made of the hardest, stiffest leather, and the heel strap was doubled at the place where it would rest on a shoe. I took another piece of moderately thick leather, perhaps two inches wide, uniting it together after the heel strap was slipped between. The job was complete, and the only instance the boot came off was when I had been negligent in renewing the hole, and the horn wore away until there was nothing to hold. The best place I have found to place the screw is about an inch outside of the center of the toe, and the leather is cut out in a curve back of the screw to near the same, and as when fastening the toe-weights and quarter-boots; the strap is buckled before it is shoved down to its place, and in this way the tugging at the strap is avoided. The job is much easier than when the other method of fastening the scalper is followed, and the danger of the strap breaking is done away with.

It is far easier to make the holes in the horn than many imagine. I use a small Morse drill, put in a handle, and a few seconds are all-sufficient. Where two screws are used there must be due care taken to drill the holes in the foot exactly corresponding with those in the metal, though this is easily done by using in the first place a Morse drill, the same size as the hole in the weight, and by giving it a turn or two the center is marked for the smaller. Drive that screw home, having been careful to give the proper angle, and then put in the other hole.

I have no desire to induce people to use any particular pattern of weight. My preferences are not offered as a guide for others to follow, and it may be that there are drawbacks that I have failed to discover.

Meager as this appendix is in suggestions, and though I am loth to present it in the shape that I do, my readers will coincide with me that it is better than to mislead with opinions which have no firmer basis than the wildest conjectures. That immeuse benefits have followed the introduction of toe-weights is beyond question; that the ill effects have been reduced at least ninety per cent. by a more intelligent use is also certain; that the advantages have far overbalanced the drawbacks I implicitly believe, and the trainer of trotters who does not make use of them in some cases is either very fortunate in the pupils he has, or is behind the times.

When colts are doing well, going squarely, improving in speed as rapidly as a reasonable man can desire, my advice is to let well enough alone. When the reverse is the case, and there is a tendency to hitch, singlefoot, pace, shorten the stride, etc., try weights, light at first, heavier if these do not correct. But it also must be borne in mind that every ounce is additional strain on the muscles and tendons, and oftentimes in the training of trotters, as in other pursuits, "the more haste the less speed."

The "side-weight" question is even more troublesome to me than that which pertains to the fore-foot, and further than to give the result of a few experiments, and the reasoning which these have led to, I will not go. Nearly twenty years ago I wrote that the action of the forelegs was more under the control of man than that of the hind, and this assumption I still hold to be correct. In common with a large majority of men of the present day who have paid a good deal of attention to the trotting action, I have changed my views, in some respect, in regard to an "open gait," and in place of considering it essential that the hind legs should be carried apart enough to clear the fore-legs on the outside, have become convinced that it is not necessary. Some of the fastest go close with their hind feet, and usually the Electioneers carry both fore and hind feet near together. If there is no cutting or wounding of coronet, pastern, shin or knee, this is manifestly a husbanding of force, the same as when there is just knee and hock action sufficient to give length of stride. Excess of action of any kind is a waste of physical force, and the smooth, easy-going trotter takes the same status as the "daisy-cutter" among race-horses. But when a horse carries the limbs so that they receive a violent blow from the passing foot, some remedy has to be discovered. Though in a great number of instances boots offer sufficient protection, there are cases when they fail to give a safeguard, and valuable as these appendages are there are cases beyond the power they possess to correct. In such a case as when the shin, ankle or pastern of the hind leg was struck, the nearly unanimous agreement among trainers was to use side-weights, under the impression that by putting them on the outside of the feet the legs would be carried farther apart. I felt the utmost confidence that this was the correct idea, and in a previous chapter give the history of a pacer that was cured of a habit of knocking his knees by weighting the outside of his front shoes and giving his slow work with an instrument in his "fork" that kept up a steady outward pressure. I am of the opinion now that the benefit came from the pressing outward, as that was continued some length of time, and when driven fast afterwards, the martingale, that part of which came between the legs, was made wide and padded.

Something over a year ago, when studying the effect of the weights, I came to the conclusion that the commonly accepted theory was wrong, and that a contrary result would follow. For instance, to take a rod, and send it swinging like a pendulum, whichever side the weight was placed, there would follow a divergence from a vertical line, the bent being away from the weight, the center of gravity would be nearer the weight in order to equalize the sides. When the leg was in motion there would be an analogy between it and the pendulum; but then arose the question of overcoming the inertia, and that might counteract the tendency to swing away from the weight. In order to give it a practical test I had a set of hind shoes made for X X, the inside web being double the outside, and consequently twice the weight. Each shoe weighed eighteen ounces, twelve on the inside from the center of the toe back, to six ounces on the outer part. Riding behind him in a skeleton wagon before the shoes were put on and afterwards, I thought that I could plainly see that he went wider. He wore these shoes from February 14th to March 30, 1882, and when pulled off that from the right foot weighed $15\frac{1}{2}$ ounces, and the left 15 ounces. This difference in the wear was, doubtless, owing to an injury to the hock joint which made it a trifle stiff, and consequently there was more of a sliding motion, and less hock action than in the other. The same day the shoes were removed I drove him to the track, and though he moved easier without the heavy hind shoes, it was perceptible that the feet were carried closer together. Before making this test I had a conversation with Hon. A. P. Whitney, of Petaluma, which arose from a question he asked. The query was: "What do you consider the proper method to follow when a horse brushes the outside of his fore-foot in passing it with the hind?" The reply was that if he had asked the question a few weeks before, I should have answered promptly: "Use side-weights on the outside of his hind feet;" but now I was in doubt of that being the right course. He then informed me that he had a horse which just grazed the fore-foot, and, following instructions, he applied side-weights. In place of "carrying him further out," as he was assured would be the result, he went so much further in the opposite direction that he struck squarely into his heel, and with so much force there as to "knock him off his feet."

The only other illustration, apart from my own trials, was a test which John A. Goldsmith made with Inca. He had the same views I formerly shared, and in a conversation last summer advocated placing the weight on the outside. One of his "string" was the stallion Inca by Woodford Mambrino, from the dam of Del Sur and Romero. He was a horse of peculiar action, twisting his fore feet outward and anything but a true-gaited trotter.

The shoes put on Inca weighed eleven ounces each, the inner side being wide enough to give twice the weight of metal, and with these there was a manifest improvement in his speed and manner of going. He had a record of 2:32, or about that, when Goldsmith got him, and after the change in his shoes he trotted at Santa Rosa in the 2:30 class, winning the first heat in 2:27, was second to Albert W in the second heat in 2:27, and won the third in 2:27¾, Albert W winning the race.

From Anteeo wearing the outside of his shoe while the inner was scarcely bright, and also thinking the three-quarter shoe so much better for the hind foot, I used that pattern.

Becoming convinced, however, that the weight on the outside was inimical, I had full hind shoes made, the outer side narrow and thick, the inner thin and wide, the object being to make the weight equal, and keep the foot as nearly level as possible during the period of wear. This did not make a perceptible difference, and so I returned to the three-quarter shoe, covering the outside in order to have a rest for the heel-strap of the scalping-boot. With the steel coming on the inside there was so little wear of the metal that in ten days or two weeks the foot would have quite a cant to the outside. This was in a measure obviated by leaving the inner side bare, and I deemed it better to guard against a wrong twist than to obtain the slight benefit of weight on the inside. But, after finding out that a scalping-boot could be kept in place without the projecting heel, I decided to put tips on behind, as the only place he wore much of the metal away was at the toe and about half-way back on the outside, and a short distance back of the toe on the inside. That part of the tip was made wider in order to equalize the weight, and they were set on December 20th. While wearing these the only work he had was jogging on the road, driven by the man who takes care of him, and he reported that he handled himself better than in anything he had worn before. On the 18th of January I pulled off the hind tips, which were worn almost to an edge, put on front tips, weighing three ounces each, and I had to cut away his heels a good deal to bring them on a level with the tip. The hind feet were left bare, and the edges rounded. I moved him through the stretch the next day, and he handled himself very well without anything in the shape of boots. Since then I have driven him occasionally, putting on toe-weights of three and three-fourth ounces, and quarter-boots, and restricting his fast work to brushes through the stretch. The hind feet I keep level by cutting the inside every few days to correspond with the wear on the outer, and being careful to round the edges to guard against breaking. When wearing tips in front and three-quarter shoes behind, all the boots he required were ankle and scalping-boots; with the tips, toe-weights and quarter-boots, and hind feet entirely bare, he requires shin boots with a speedy-cut attachment, and these he marks with the outside of the fore-foot, and occasionally brushes the outside of the quarter-boot with the inside of the hind-foot. Now what is the cause of this going wider behind I will not undertake to say with any degree of authority. I think he is showing more speed than he did previously, as he hauled the cart through the stretch in thirty-eight seconds, and I only drove him at his best for about two hundred yards. Last summer the toe-weights were detrimental, retarding his speed in the few trials I gave them. Now they are surely an advantage, or else it is the change in his disposition that must be credited with the improvement. As he has become less obstinate, more ready to go and a willingness to trot from the start, there may be something more potent than the weights to ascribe the change to. At all events, the data is presented, and my readers can draw the inferences.

INDEX.

CHAPTER I.—NECESSITY FOR A BETTER SYSTEM OF SHOEING.
Guards against concussion—Breaking the jar—"Round-hoof'd"—Dilatation and contraction—Weight controls the action—Tight boots.

CHAPTER II.—GUARDS AGAINST CONCUSSION—AN ELASTIC SHOE—STONEHENGE ON TIPS, ETC.
Artificial requirements—Unfettered quarters—The English greyhound—The Oakland streets—"Dwelling" action.

CHAPTER III.—DIFFERENT FORMS OF TIPS—EFFECT OF WEIGHT ON THE FEET—MILES' ESSAY—ETC.
Best method of nailing—The rolling-motion shoe—Changing the pace into the trot—Open-heeled shoes—An admirable contrivance.

CHAPTER IV.—CURE OF A SPRUNG TENDON—RESULTS OF EXPERIMENTS—ETC.
A surprising effect—Hock-Hocking.

CHAPTER V.—FROM SHOES TO TIPS—FURTHER SATISFACTORY TESTS.
Intelligent trainers—One-sided tip—An important subject.

CHAPTER VI.—MISTAKES OF BLACKSMITHS AND GROOMS—HARD ROADS AND THE CONSEQUENCES.
A singular influence—Practices of the vaquero—Corns and contraction—Elasticity of the sole.

CHAPTER VII.—DIFFERENCES OF OPINION IN REGARD TO PROPER ACTION—EFFECTS OF CHANGES IN SHOEING—PRACTICAL EXAMPLES, ETC.
Colonel Lewis and Fullerton—Effects of a heavy shoe—The trials of Avola—Restricted Experiments—Contrary effects of weight.

CHAPTER VIII.—ENDORSEMENT OF TIPS—THREE-QUARTER SHOE.
Hampering the Growth—Wear on the outside—Correcting faulty action—The Goodenough shoe—Decreased strain on the tendon—An effective friction-break.

CHAPTER IX.—GUARDS AGAINST CONCUSSION.
Foot of the elephant—Rational ox-shoeing—Erroneous setting of tips—Stride of the race-horse.

CHAPTER X—GROWTH OF THE HORN—PROTECTION TO THE FOOT.
Elasticity of natural horn—Bedding the tips—A practical illustration—A convert.

CHAPTER XI.—A NATURAL FOOT VS. A PERFECT FOOT.
Cracked heels—Lady Viva.

CHAPTER XI (*Continued*).—REASON FOR WANT OF CONNECTION—HISTORY OF ANTEEO.
The first Embryo—Width and elasticity of frog.

CHAPTER XII.—HISTORY OF ANTEEO CONTINUED—SKETCH OF ANTEVOLO.
An incorrigible—A trial with toe-weights—Regeneration—A projecting tip—Antithetons.

CHAPTER XIII.—ENGLISH IDEAS OF HORSE-SHOEING—TIPS AND CHARLIER SHOES.
Tips on saddle-horses—Tips on hunters—Error in paring the sole—Infinitesimal tips—Motives for shoeing—Necessity for nailing—Work with Nature—Miserable life, premature death—Tips to be tried—Still experimenting.

www.ingramcontent.com/pod-product-compliance
Lightning Source LLC
Chambersburg PA
CBHW020906230426
43666CB00008B/1327